T0342443

Shellac in Visual and Sonic Culture

For Brigitte, William and Benjamin Roy

Shellac in Visual and Sonic Culture

Unsettled Matter

Elodie A. Roy

Amsterdam University Press

Cover illustration: *Etude publicitaire pour Magic Phono, portrait de Marie Bell en photomontage, François Kollar (1904–1979)*, 1929. Charenton-le-Pont, Médiathèque du patrimoine et de la photographie (C) Ministère de la Culture – Médiathèque du patrimoine et de la photographie, Dist. RMN-Grand Palais / François Kollar (C) RMN – Gestion droit d'auteur François Kollar.

Cover design: Coördesign, Leiden
Lay-out: Crius Group, Hulshout

ISBN	978 94 6372 954 3
e-ISBN	978 90 4855 314 3
DOI	10.5117/9789463729543
NUR	670

© E.A. Roy / Amsterdam University Press B.V., Amsterdam 2023

All rights reserved. Without limiting the rights under copyright reserved above, no part of this book may be reproduced, stored in or introduced into a retrieval system, or transmitted, in any form or by any means (electronic, mechanical, photocopying, recording or otherwise) without the written permission of both the copyright owner and the author of the book.

Every effort has been made to obtain permission to use all copyrighted illustrations reproduced in this book. Nonetheless, whosoever believes to have rights to this material is advised to contact the publisher.

Table of Contents

List of illustrations

Acknowledgements

This book began its life in 2016–2017 at the Department of Cultural History and Theory at the Humboldt University of Berlin. I am grateful to all my former colleagues in Berlin – and particularly Holger Brohm, Britta Lange, Ricardo Cedeño Montaña and Tiago da Costa – for being so supportive in the very early stages of this project. I would like to thank the colleagues who further encouraged me and invited me to write about shellac in the past years: Andrea Bohlman, Alexandrine Boudreault-Fournier, Kyle Devine, Fanny Gribenski, Petra Löffler, Antonio López, Jonathan Hicks, Antje Krause-Wahl, David Pantalony, Peter McMurray, Eva Moreda Rodríguez, Änne Söll, Viktoria Tkaczyk and Christine von Oertzen.

I would also like to thank Carolyn Birdsall, Sean Dodds and Frances Robertson for attentively reading sections of this book and offering valuable comments, the two anonymous readers for their constructive feedback on the first version of the manuscript, and Maryse Elliott at AUP for commissioning this book.

I would like to acknowledge and thank here the artists and designers who shared their passion for shellac and accepted to be interviewed: Dinah Bird, Sascha Brosamer, Jim Dine, Christopher Dorsett, Graham Dunning, Anja Lapatsch, Robert Millis, Irene Pérez Hernández, and Annika Unger.

The Zinsser and Walworth families – particularly John Zinsser, Seth Walworth and Cornelia Walworth – have also been supportive of the project and I would like to thank them here.

A grant from the Association for Recorded Sound Collections (ARSC) allowed me to pursue archival research at the British National Archives and the British Library in 2019. This research further benefitted from the support of the Center for Material Culture Studies at the University of Delaware. I am grateful to them for recognising the value of this project.

Finally, my heartfelt thanks go to my friends Benjamin Belinska, Ian Helliwell, Sophie Robinson, Jamie Sexton, Adam Schofield, João Silva and Karin Weissenbrunner.

This book is dedicated with my deepest love to Brigitte, William and Benjamin 'Pidou' Roy who were there at the beginning and have supported me unconditionally over the years.

Introduction: From material culture to the materials of culture

Abstract:

The introductory chapter surveys the main theoretical concerns and themes underpinning the book as well as giving a brief historical overview of what shellac is. It gives insights into the chosen methodological framework, surveying what the implications of following the mutable materials – and stories – of media cultures are. Rather than focusing on finite media objects and practices of consumption, the introduction highlights what an emphasis on materials and processes might mean for the study of media cultures.

Keywords: media archaeology, materiality, ecology, narrativity, shellac

In the Spring of 1936, the London Shellac Research Bureau celebrated shellac at its India House headquarters in Central London.[1] One of the windows of India House offered a modest yet carefully curated display of shellac-based artefacts (see Figure 1 below). The material, a thermoplastic of insect origin imported from British India, appeared in its various sizes and guises, ranging from tiny soluble flakes held in fragile glass vials to finished commodities such as bowls and hats.[2] Among the everyday objects featured in the window display, gramophone discs – perhaps the most iconic and best-known of all shellac-based artefacts – occupied a prominent position. Emile Berliner's shellac discs had been introduced in the second half of the 1890s. By the 1930s, as is still the case today, shellac was principally and most spontaneously

[1] The London Shellac Research Bureau, which was attached to the British Government, worked in close connection with the Indian Lac Research Institute in Ranchi (at the heart of the lac-production area) and the Shellac Research Bureau in New York. In London, India House hosted the Bureau from 1934 to 1940, when the Bureau was transferred to the University of Edinburgh for fear of German bombings.

[2] It is the only known plastic of animal origin.

Roy, E.A., *Shellac in Visual and Sonic Culture: Unsettled Matter.* Amsterdam: Amsterdam University Press, 2023

DOI 10.5117/9789463729543_INTRO

Figure 1: Shellac window display, India House, London, 1936. Author's picture. National Archives, UK.

associated with audiophilic pleasures and the enticing black sheen of records, more closely relating to the realm of sound rather than to vision. In 1935, half of the shellac that England imported was to manufacture gramophone records and the material, which had previously been consumed by 'the highly industrialised countries of Europe and America', now attracted new clients such as Japan and Russia.[3] India was the first – and almost sole – worldwide producer of shellac before the Second World War, providing 90% of the global supply and processing the remainder.[4] At the time, the multinational gramophone industry represented the main single consumer of shellac in the world, absorbing over one third of the annual output.[5] Shellac cultivation was almost exclusively concentrated in the region of Bihar, with smaller production areas in the United Provinces, Bengal and the Central Provinces.[6]

3   See Parry (1935), 170, and Adarkar (1945), 2. Russia had a developing record industry and production programme: the country, which produced 700,000 records a year in 1934, planned to manufacture 40,000,000 records annually by 1937 (the equivalent of 2,000 tonnes of shellac a year).
4   Spate (1964 [1954]), 233. In addition to this, lac products alone constituted almost half of the total value of Indian forest products (Ibid.).
5   Adarkar (1945), 1. The early record industry constituted one of the first truly transnational business. The Gramophone Company (to become EMI in the 1930s) was established in 1898. In the first decade of the twentieth century, a number of commercial recording expeditions were undertaken in India, Japan and China, to sonically capture – and ultimately commoditise – colonial and subaltern subjects. See Jones (1985).
6   Bihar was also, incidentally, the centre of mica production – another strategic material for the electrical and radiophonic industries. See Adarkar (1945), 3.

Rather than concealing it, the exhibit at India House celebrated the longer and larger history of the British Empire. A large photograph showing a small, whitewashed Indian shellac workshop was hung in the background, hovering about the familiar assemblage of everyday objects. The uncaptioned photograph pointed to the often-ignored origins of shellac production, visually connecting the early record-making industry to a long, global quest for resources. The name 'India House' itself concretely evoked the powerful East India Company of the early seventeenth century which had first imported shellac into England. The original India House, located in the Leadenhall area of London, had been the place where colonial goods such as shellac, but also tea, cotton, silk, tin and spices, had been auctioned off to be disseminated across Britain from the seventeenth to the nineteenth century. The 1936 shellac exhibit at India House was one of many events organised to promote shellac-based commodities throughout the 1930s. The polyvalent natural plastic was also exhibited by the British Government on a number of much larger national and international events. These included the 1934 Birmingham Trade Exhibition, the 1934 Canadian National Exhibition (in Toronto), or yet again the 1935 Universal and International Exhibition (in Brussels) – as well as the 1939 British International Fair (held jointly in Birmingham and London).

With its dense and heterogeneous array of objects, the window display shared affinities with the inventories of the everyday so minutely assembled by Georges Perec in his 1960s and 1970s novels (including the monumental *Life: A User's Manual*, published in 1978). Perec attempted to completely 'exhaust' everyday spaces – from streets, buildings, flats, and rooms to office drawers – by indexing and describing every item they contained.[7] Perec's inventories proposed a different mode of storytelling – operating in space rather than in time, and fusing description with narration, still life with movement. His forays into the ordinary surfaces, affects, and objects of everyday life were later theorised in his literary hybrid *L'Infra-ordinaire* ('The Infra-ordinary'). The book, posthumously published in 1989, encouraged readers:

> to question what seems to be so obvious that we have forgotten its provenance. To recover something of the puzzled sensation that Jules

---

7   Visual artist Daniel Spoerri, in his three-dimensional collages or 'picture-traps' spanning the same period, would similarly draw cartographies of the everyday by gluing together collections of vernacular objects. In the early hours of the morning on October 17, 1961, Spoerri resolutely traced the objects cluttering his studio table, making a map of them. In the months to come, every artefact would be numbered, captioned and described by the artist, its history tentatively retraced, its meaning momentarily circumscribed yet never fully suspended or exhausted. See Schwenger (2006), 101.

Verne or his readers felt in front of a device that could reproduce and transport sounds. [...] We must question bricks, concrete, glass, table manners, devices, tools, timetables, rhythms. Question that which has now ceased to puzzle us. [...] How? Where? When? Why?[8]

Perec's prompts remain useful. Drawing on his insights, I would like to suggest that a similar attempt can be made at unpacking the composite materialities, histories, temporalities and socialities of the artefacts photographically frozen behind the glass panels of India House. Taken together, the exhibited objects provided a deceptively arrested image of the everyday (or a version of it); on an individual level, each of them discretely indexed the ambiguous history of a single yet mutable material of culture.

Plastic stories and processes

This book is not only concerned with discrete (media) cultural objects – such as the ordinary gramophone records displayed at India House in 1936 – but also with the material or 'stuff' of which such everyday objects are more invisibly made. It argues that the investigation of processes and materials themselves – which are often dismissed as being mundane, secondary or indifferent – yields important insights into the understanding of cultural practices and epochs. Contrary to rare and luxurious materials – including metals such as gold and silver –, shellac cannot be taxonomically described as a noble substance: as such, it has never been pursued or construed as a singular 'object of desire' (to adopt Adrian Forty's formula) in and of itself, though it gradually became valued across centuries for its visual, plastic and sonic properties, as they were revealed when it was combined with other substances.

The story of shellac appears as a story of displacements, mutations and interruptions – it is also an open-ended story of intersensory recycling and rewriting. As the India House display didactically showed, the 'exotic' thermoplastic substance lent itself to myriad processes of transformation and reinterpretation – whilst fashioning and transforming, at the same time, the heterogeneous subjects who handled it (from shellac factory workers to record listeners). As we will see throughout the book, shellac results from numerous manual, machinic as well as symbolic and cultural processes of association and transformation. It is not magically given in

8   Perec (1989), 12; my translation.

or by nature: as such, there is no 'natural' or 'raw' material to speak about. The production of shellac is driven by intense patterns of technical and social processing. As well as relying on extensive human labour, its global circulation up until the mid-twentieth century would not have been possible without the endurance of significant historical, material and ideological infrastructures – most notably the maritime trade routes between India, Europe and North America –, directly inherited from the colonial expansionism of the seventeenth century. Reflecting upon the circulation of art materials, art historians Christy Anderson, Anne Dunlop and Pamela H. Smith suggest that

> [m]aterials are enmeshed with the physical as well as the societal structures of any age, and constantly in motion. Materials have a history that may be charted over the short and long term. The story of materials begins with the story of matter and matter is a product of the earth with its geological time frame. One way to think about the cultural logics of materials, therefore, is to see them as part of historical epochs and central to the social structures with which they engage. This dynamic approach links human institutions such as cities, trade networks and linguistic systems with a slowly evolving material world.[9]

While one of the aims of the book is to defamiliarise media objects by addressing their material substrates and surfaces, my intention is not to desocialise or autonomise artefacts but to assess how materiality may function as a 'substrum of the social' while being irreducible to it.[10] The total defamiliarisation of everyday artefacts – a project which was most strikingly undertaken in the early twentieth century by the Surrealists – may never be realistically achieved, for it may be that 'once we ourselves have become socialized, we can no longer see objects in their raw and unprocessed state'.[11] Indeed, while the Surrealists – including writers such as Robert Desnos, André Breton, Louis Aragon and Philippe Soupault – sought to defamiliarise the everyday through the medium of the imagination, Georges Perec hyperbolised the real through his intensive and exhaustive scrutinisation of the surfaces of everyday life. His meticulous, hyper-real descriptions paradoxically contributed to derealising lived spaces – opening up myriad of hidden worlds within the deceptively stable world of objects.

9   Anderson, Dunlop, Smith (2015), 12.
10   Ibid., 4.
11   We may ask whether such a state exists. See Thompson (1979), 77.

This study therefore acknowledges that scholarly writing, too, by reclaiming apparently innocuous materials as well as recovering seemingly anodyne gestures, overlooked processes and buried discourses, may achieve a certain degree of defamiliarisation and distanciation. For the present purposes, defamiliarisation should be understood as an attempt to defamiliarise our gaze, and an invitation to momentarily leave aside ossified discourses and beliefs to draw together fresh sets of connections.

What follows provides a conceptual and methodological framework for the book, introducing the main themes and concepts informing it (including ideas of plasticity, narrativity, intersensoriality, and the everyday). It contextualises *Shellac in Visual and Sonic Culture* within the existing literature on media and material culture theory – while also signalling breaks from existing theories and introducing alternative patterns and points of departure. This book is part of a much larger continuum of works addressing issues of (eco)materiality. There are many reasons why contemporary scholars – across a wide range of disciplines – should be so preoccupied with the question of materials and media supply chains: the global environmental crisis, the exhaustion of resources, the pursuit of ecologically and economically sustainable alternatives may all be seen as significant factors in fuelling this (frequently anguished) interest in the material world. On an everyday level, the disquieting feeling of 'losing touch' with the real may prompt individuals to reconnect with tangible objects and physically reconnect with ritualised practices of engaging with the world (such as record-listening). It may be that material artefacts help mitigate feelings of alienation and distance – they help slow down, stabilise, and perhaps even organise everyday life. There is a comfort in familiar things, a physical immediacy which is often lacking in digital mediations. Yet digital culture and its ubiquitous totem – the smartphone – is as much a culture of distance as it is a culture of touch, connectivity and hyper-tactility (though it is rarely, as is becoming increasingly and painfully apparent, a culture of contact). Our so-called dematerialised environment is overwhelmingly tangible, with the distribution of data relying on large-scale and very concrete logistical networks.[12]

While ecomaterial thinking tends to focus on (and depart from) the present moment, this book engages with a historical media material. I believe that historical colonial resources such as shellac prompt us to reconsider the imperialistic and extractive underpinnings of contemporary media supply chains. However, while shellac is a politically charged subject matter,

12   See Schmidgen (2022); Hockenberry, Starosielski and Zieger, eds. (2021).

I also believe that reducing it to a political symbol or mere synecdoche of the colonial regime is far too restrictive. This is why this book, combining empirical and theoretical elements, adopts a wider angle. It interweaves – without systematically hierarchising them – a range of perspectives and moments in the long history of the medium. It addresses the undeniably repressive ideology informing the shellac trade network, its toxicity and effective destructivity *alongside* the intersensory culture of expressivity and creativity it gave rise to. It must be noted that there is something at once timely and irreparably anachronistic about shellac. On the one hand, it clearly is an obsolete mediatic resource – one which apparently bears no direct relevance to the digital present. In everyday conversations, the word itself is more likely to refer to the cosmopolitan world of nail salons and beauty parlours than to the commercial speculations of the East India Company or the patient experiments of Emile Berliner. On the other hand, while shellac indisputably became a devalued mediatic resource in the aftermath of the Second World War, I believe it remains a particularly potent resource for reactivating (and transmitting) stories about past and contemporary media networks – allowing us to return to the present moment with refreshed insights and a heightened sensitivity to its issues. Eventually, I see this study as an invitation to listen differently (in the wider sense of the term) and a means of opening a different reflective space – if only for a moment. It also constitutes a deliberately anachronistic move, with shellac and its histories (some of them long discarded) providing an oblique entry point into contemporary media culture.

Plasticity and ecology

Throughout the book, I am interested in what remains too often unacknowledged in media theory: the productive materiality of mediating substrates, and the ways in which this materiality itself complicates, transforms (or even defeats) the possibility of mediation. The discrete cultural logics of materials cannot be dissociated from the dynamic *politics* of materials – what Jane Bennett terms their 'vibrancy' or 'vitality'–, that is to say from the larger social and symbolic networks into which they enter and which, conversely, alter them.[13] For Bennett and theorists of vibrant matter, no primacy of form over meaning may be assumed. Rather than a definitive account of shellac production or an exhaustive inventory of its uses, what interests me most is

13   See Bennett (2010).

precisely its capacity to become something else or contribute to new media assemblages and, by extension, networks of meaning – across extended stretches of time and space. Shellac is literally plastic, in the full sense of the adjective. The term 'plasticity' (from the Greek *plassein*, 'to model' or 'to mould') both evokes the (passive) 'capacity to receive form', and the active 'ability to give form to something' – or to annihilate form (as in the case of plastic explosives).[14] It follows that what we call 'shellac' – a deceptively homogenising term – both covers and generates a variety of material realities across eras. Rather than suggesting that materials are strictly timeless or transhistorical – as implied in the traditional distinction between the 'brute intransigence of matter, everywhere and always the same' and the 'plasticity of meaning, bound to specific times and places'[15]–, the present book argues that matter itself may be historical, historicised, and mutable. No hard boundaries are posited between 'matter' and 'meaning'. Rather, bearing in mind Karen Barad's definition of 'mattering' as 'simultaneously a matter of substance and significance', symbolic and material practices are understood as co-emergent.[16]

Throughout, shellac is understood ecologically –, where '[e]cology is a science of relations and mediations, in which innumerable interactions must constantly re-create the end points "environment" and "inhabitant"'.[17] It follows that media can further be 'defined as assembled of various bodies interacting, of intensive relations. Media can be seen as an assemblage of various forces, from human potential to technological interactions and powers to economic forces at play, experimental aesthetic forces, conceptual philosophical modulations'.[18] What Parikka highlights here, in a passage inspired by Deleuze and Guattari's assemblage theory, is the open, unstable nature of (media) assemblages which are continuously forming, unfolding, and falling apart. Bodies, here, are widely understood as being physical, institutional, personal, collective, industrial, scientific, and so on. Assemblages, to the extent that they are 'constituted by a relationality',

---

14   Crockett (2010), xiii. In recent years, the terms 'plastic' and 'plasticity' have been redeemed in the field of philosophy, particularly through the extensive work of Catherine Malabou (2010, 2012). Malabou uses a very specific concept of 'plasticity' (first theorised in the preface of Hegel's *Phenomenology of Spirit*) in relation to the continuous forming and annihilation of human subjectivity: plasticity, she writes, 'can signify both the achievement of presence and its deflagration, its emergence and its explosion' (Malabou 2010, 8).

15   Daston (2004), 17.

16   Barad (2017), 3.

17   Cubitt (2017), 9. On media ecologies, see also Fuller (2005) and Herzogenrath (2015).

18   Parikka (2010), xxvi.

are always already precarious, situated and temporary.[19] The present study takes up this focus on ecology and co-dependency, in anticipating a more holistic and intersensory approach to media objects. As such, it leads to a politically aware approach to the concept of recorded sound, positing the building of phonographic cultures at the turn of the twentieth century as a horizontal and transnational process, relying on a wealth of frequently anonymous intermediaries and materialities. My understanding of phonography therefore bears in mind the localised historical practices, ideologies, and techniques involved in the production of musical commodities.

Importantly, the ecological perspective regards materials as extensive, temporal, changing and agential rather than eternally immutable. Material culture scholar Fernando Domínguez Rubio further argues that to think ecologically may first and foremost mean to recognise the precariousness of the material world as well as the continuous processes of care and maintenance by which it is physically and symbolically stabilised.[20] Cultural objects, in order to retain their legibility as cultural inscriptions, must be incessantly repaired – both in the physical sense of the term and, we may add, in the sense of a symbolic reinterpretation or reactivation. Conservation is therefore predicated upon transformation, where the latter retrospectively and continuously 're-creates' (or re-forms) the appearance of the past.

Narrative assemblages

Materiality cannot be separated from narrativity and processes of storytelling, conceived of as a three-dimensional practice akin to weaving or sewing together heterogeneous elements – though not in a random manner. Indeed, we may only understand materials when we begin 'tell[ing] their histories – [...] what they do and what happens to them when treated in particular ways – in the very practice of working with them'.[21] To some extent, storytelling – and the telling of histories – may therefore be conceived of as a reparative mode – one which doesn't seek to create a totalising whole or unity, but which paradoxically combats fragmentation by magnifying and incorporating it (as it draws attention to visible stitches, sutures, disparities and disconnections). Media theorist Sean Cubitt productively suggests that stories and anecdotes can be recuperated as a heuristic method and terrain

19 Ibid., xxv.
20 See Domínguez Rubio (2016).
21 Ingold (2012), 434.

of experimentation – forming a ground which *precedes* theory but without which the work of theorisation would not be possible.[22] In my reading, Cubitt's plea for the recognition of 'anecdotal evidence' is less an invitation to uncritically 'incorporate' the anecdote into an otherwise stereotypical narrative form than an invitation to revise – or, at least, re-examine – the conditions and structures of knowledge transmission itself. As such, the study of anecdotes also anticipates an active interrogation and revaluation of modes of (historiographical) writing.

In his 1979 book *Rubbish Theory: The Creation and Destruction of Value*, surveying the cultural and economic revaluation of outmoded artefacts, Michael Thompson already proposed that a rhapsodic mode of writing may help us represent and better understand processes of symbolic and physical transformation (within what he calls the field of 'rubbish theory'). He urged scholars of the material world 'to deal in different forms of discourse simultaneously. And since they cannot be mixed they must be juxtaposed. The joke, the paradox, the shock technique and the journalistic style, far from being unscholarly devices to be avoided at all costs, become rubbish theory's inseparable accompaniments'.[23] The present book combines various (and sometimes jarring) voices and modes of writing – it includes sources as diverse as extracts from seventeenth-century treatises and travel diaries, excerpts from twentieth-century sound recording manuals, anecdotes and mythical accounts. I believe that stories may be understood as a form of waste or unassimilable surplus, pointing to the 'malleable multiplicities of the world' and its open-ended becoming.[24] The rich, embodied plasticity of anecdotes may be opposed to the decontextualised rigidity of data, with the former being 'not things but actions, which is why they appear so often in the form of stories'.[25] Within Cubitt's larger ecocritical project of pursuing the 'good life', the anecdote is strategically mobilised as 'the unique instance that reveals the forces operating and the possibility of their working otherwise'.[26] While this book is not a strict ecocritique in Cubitt's sense, it embraces the plastic heterogeneity of stories and their potential to '[unpick] the stability of the given, the fait accompli'.[27] I argue that the anecdotal methodology becomes especially relevant when it is used in relation to the composite media cultures of the early modern world, as it

22 See Cubitt (2013); Cubitt (2020).
23 Thompson (1979), 5.
24 Cubitt (2020), 35.
25 Ibid., 2.
26 Ibid., 5.
27 Ibid., 14.

allows us to address and better understand their built-in heterogeneity and unfinishedness.

An apparently narrow starting point such as 'shellac' (or any other material) therefore invites to a process of expansion or unfolding which could be compared to a diffractive or reticular approach. How does one engage with the multi-sited and multi-temporal, fragmented 'histories of materials'? A complementary question would be: How is it possible to shift from a logic of rigid, linear representation to one which embraces the unsettled and performative nature of materials? How does one study materials and their vibrant processuality? Art historian James Elkins once gloomily suggested that, since no coherent theorisation of materials was possible, art history should be content with giving us a general theorisation of 'materiality'.[28] Yet I believe no unique or unifying theory is quite able to accommodate the fluidity of materials: perhaps this calls for a flexible or modular theory, a theory of materials in motion which doesn't flatten, detemporalise or dehistoricise materialities (as Actor–Network Theory often does for instance), but which acknowledges the differential materialities and temporalities of things.

The question of writing – and especially of writing about what keeps changing – is intimately linked with the genesis of material culture studies and continues to represent a key question in the field. For material culture theorists, the study of materiality cannot be divorced from a close examination of what it means to narrate materiality – and, on a practical level, to 'translate' matter into language. In a seminal essay first published in 1964, John A. Kouwenhoven, studying early American material culture, sought to coin a new mode of writing to accommodate sensory affects, anticipating the methodological issues which would be central to the field of sensory studies at the turn of the twenty-first century – and particularly in multisensory and multidimensional approaches to film and media.[29] Like Kouwenhoven, I believe that there exists an intermediary zone between the deceptively 'raw' material and the 'polished' theory or historical narrative – and I'm interested in storytelling as a material practice. Accordingly, the present study is concerned with shellac – and media objects more broadly – in their full intersensory and embodied (or carnal) dimension. It is also concerned with experimental ways of writing about (and understanding) them – and with the work of communities (particularly, but not only, creative practitioners) engaged in physically

28   Elkins in Lehmann (2015), 25.
29   Kouwenhoven (1982), 88. See also Marks (2000); Sobchack (2004); Schmidgen (2022).

reworking and transforming the material and its meaning (as explored in Chapter 5).

My interest in shellac lies with the plasticity of representation as much as it does with the physical plasticity of the material itself. It may be suggested that writing about the materiality of shellac can become an experiment with the materiality of writing, where matter gets transformed in the act of retelling. Reciprocally, stories can be understood as heterogeneous material artefacts: they are incessantly crafted and assembled, passed on, discarded, and recast – though never in a strictly linear way. Rather, the 'shape' of a story may precisely be the story of its deformations, the synchronous record of its scars. Much like the actual shape of a house, which records the passage of those who have inhabited it, the story bears the imprints of the hands it has passed through – in Benjamin's poetic words, 'the handprints of the potter cling to the clay vessel'.[30] The marked, marred surface of the story thus appears to be inseparably bound with its symbolic contents; surface and depth coincide. It is no surprise that the potter, intimately working with and through the grainy resistance or compliance of clay, should be so central to Walter Benjamin's understanding of storytelling as a three-dimensional, nonlinear and nomadic practice. The image of the weaver, too, would be equally appropriate to describe the storyteller. There is an itinerant, repetitive quality inherent to the art of telling stories, which presupposes a discipline and a routine – yet the repeated gesture of the storyteller, leaning as she does on familiar, time-worn images and narrative devices, never excludes an unforeseen flight of fantasy, or a spontaneous leap into unchartered and uncanvassed territory. On the contrary, it may be proposed that discipline and repetition patiently prepare and authorise departures from the known grid of the narrative. In storytelling as well as in weaving, the entwinement of patterns cumulatively composes the familiar backdrop against which novel knots of meanings may be fastened.

The present book engages in a process of remediation, understood in the broadest sense of the term as a reparative mode.[31] Such a practice of

---

30   Benjamin (1973), 92. Benjamin, writing in the aftermath of the First World War, was concerned with the disappearing craft of storytelling in industrial modernity, characterised by novel and (for Benjamin) often impoverished modes of communication. He notably saw this decline as related to the devaluation of experience and the rise of merely instrumental, ephemeral, mass-produced information whose meaning, he wrote, could 'not survive the moment in which it was new' (Ibid., 90). Paradoxically, as he lamented the passing of the storytelling age, Benjamin became in turn a storyteller, a mediator and a redeemer of history in its vertiginous, ever-generative fullness (opposing the empty temporal shell of the 'now' and the 'new').

31   See Bolter and Grusin (1999).

re-mediating media 'stories' and 'anecdotes' differs from an automatic and empty retelling of what media archaeologist Wolfgang Ernst calls 'media-historical narratives', with their false linearity, consistency and stability.[32] It may be briefly recalled here that Ernst's aim is not simply to deconstruct the grand narrative of media-historical writing: more radically, his is an attempt to go *beyond* (or overcome) 'alphabetic' writing itself, by focusing on discrete machinic temporalities and adopting alternative modes of description (mostly mathematical) stemming from (digital) technologies themselves. Thus, he invites us to dispassionately 'count' rather than 'recount', and to become machine-like, whilst gently conceding that 'the two methods [*counting and recounting*] will continue to supplement each other without effacing their differences in *parallel lines*'.[33] As it seeks to achieve a non-phenomenological and non-narrative mode of understanding media predicated upon emotional suspension and withdrawal, radical media archaeology continues to profess a deeply rooted suspicion of human subjectivity. The difficult non-narrative (or anti-narrative) methodological path outlined by its proponents – liberated from the compulsion of historical retelling or projection –, effectively sheds a different, colder light upon media artefacts. For the point is to understand media *from 'the perspective of the media themselves'*, and learn their own mathematical language.[34] The allure of radical media archaeology may lie precisely in its efforts to attain the impossible.

My method here is much closer to the human logic of recounting rather than that of machinic counting. I do not equate storytelling with the totalising, seamless linearity typically associated with the term 'grand narrative'. As such, rather than a masking or homogenising operation – where homogeneity may only betray an alienated form of discourse –, I would like to propose that storytelling may constitute an essentially disrupting and decentring mode of (re)presenting and producing knowledge, as it draws attention to (rather than concealing) the links which exist between disparate elements of the narrative whole. Of course, the links themselves are never 'given' or *sui generis*: they are partially – and always cautiously – produced by the act (and actuality) of juxtaposition. Juxtaposition mimetically retains the literal grain – or archival quality – of material culture research, characterised by ruptures and contingencies. In this respect, the experimental, nonlinear story-telling approach closely resonates – in a roundabout way – with media

32   Ernst (2013), 55.
33   Ibid., 71; 54.
34   Ernst (2011), 240; emphasis is mine.

archaeology's core concerns with epistemic, spatiotemporal, and material discontinuities. It also aligns with the discipline's habit of self-scrutiny (or energetic re-tracing of one's steps).

Everyday materials

Design historian Henry Petroski famously charted a whole history of writing through a study of the pencil, suggesting that 'to scrutinize the trivial can be to discover the monumental. Almost any object can serve to unveil the mysteries of engineering and its relation to art, business, and all other aspects of our culture'.[35] There are similarities to be found between the material-driven approach of media archaeology – and its commitment to practices of retro-engineering – and some strands of material culture studies embracing a multi-scalar approach encompassing the infinitely small and the incommensurable, the miniature and the monument.[36] It follows that discrete and apparently trivial everyday objects such as gramophone records may become composite gates into global infrastructures – helping us navigate the myriad scales and materialities of capitalism.[37] Moreover, it may be that the focus on (dynamic) materials of culture allows us to draw a finer, differentiated model attuned to both the evential and the repetitive (or the cyclical), the local and the global, the short and the long term, the seasonal and the historical.[38] On a practical level, this means that at least two main narrative modes – the seasonal and the historical (with their distinct registers, rhythms and paces) – must be woven together. It is because of its seemingly amorphous elusiveness, as Henri Lefebvre suggests, that the everyday cannot be easily written about or archived, thus escaping processes of historical mediation and recuperation. However, the materiality of the everyday – and its intimate reliance on objects and infrastructures – is precisely that which may allow us to reflectively 're-historicise' it. And yet such recuperation is not straightforward or total: to historicise the everyday also means, paradoxically, to embrace its shapelessness and recognise its indefatigable state of becoming. Accordingly, to look at materials in motion may be a means of defetishising

35 Petroski (1989), 27.
36 See Petroski (1989); Asendorf (1993); Stewart (2007 [1993]).
37 Esther Leslie's recent theoretical forays into the micromateriality of dust, for instance, reconnect it to the building (and crumbling) of historical epochs. See Leslie (2020).
38 See Lefebvre (2013 [1992]), 18. See also Kubler (1962) on the interrelation of different 'shapes of time'.

mainstream material culture studies and their typical fixation on singular or iconic media-cultural objects and practices of consumption.[39] In such an endeavour, we must remain careful not to replace the cult of objects with a cult of materials – or to fashion new totems out of discarded materialities. It is precisely the emphasis on processes of cross-cultural and media-material transformation which may allow us to avoid the dangers of petrifying the subject matter, and to better attend to the incessant transformations and deformations of media cultures.

In addition to this, modest materials, though they are not objects per se, constitute artefacts in their own right: crucially, they are able to mix with other materials and to become something else in the process of association. Throughout, I consider shellac as an 'informed material' which is 'transformed through [its] changing associations'[40] – by analogy with the chemical realm where atoms and molecules ceaselessly acquire different properties and identities depending on their environments. A focus on changing materials, forms of labour and processes allows us to ask different questions about media-material environments – bringing to light their pre-mediatic life as well as their afterlives. To think about materials is always already to think about matters of heterogeneity, transformability and temporality. A focus on discrete media materials such as mica, steel or shellac dissolves the 'hard' or stable objects of technological history. In media archaeology, the enquiry shifts from machines and models to discrete materials and processes. When media history gets broken apart, it is reassembled and recycled as 'one big story of experimenting with different materials from glass plates to chemicals, from selenium to coltan, from dilute sulphuric acid to shellac, silk and gutta percha, to processes such as crystallization, ionization, and so forth'.[41] The large-scale, interconnected 'big story' does not replicate (and should not be confused with) the 'grand narrative' of media history: it is best understood as a story of multitudes, a swarming or a 'relational whole' which can never be fully resolved as a unified, or uniform, entity.[42] As such, it only ever assumes a provisional shape. However, I would argue that rhapsodic practices of assembling media knowledge are not antithetical to narrative logic but, rather, generate their own mode of narrativity and of heuristic efficiency.

39 This is what Ingold calls the 'stopped-up objects' of mainstream material culture as opposed to the 'leaky things' embraced by ecomaterial approaches; Ingold (2012), 438.

40 Barry (2005), 57. See also Westermann (2013), 81.

41 Parikka (2012), 97.

42 Parikka (2010), 47.

Decentring media

An emphasis on materials and processes ultimately invites us to return to cultural objects (rather than completely abandoning them) – including artworks and media artefacts – in relation to '[their] conditions of production, their physical creation out of the materials of the earth'.[43] A narrow focus on finished media objects, machines and commodities often leads to overemphasise their industrial histories and histories of consumption (such as record-listening), therefore drastically limiting or silencing what can be said about their conditions of emergence. While it is not possible to minimise Emile Berliner's crucial contribution to the development of the gramophone disc (of which Chapter 2 gives an account), it may be argued that the story of the gramophone starts long before Berliner's discovery of the sonic properties of shellac in the mid-1890s – and does not stop with the invention and commercial dissemination of the artefact. In Parikka's words, the 'materiality of media starts much *before media become media*', urging us to '[find] strains of media materialism outside the usual definition of media' in order to address the hidden substrate of media cultures (notably matters of labour).[44]

In the case of phonography, studies of the social life of the gramophone record tend to be restricted to the consumers and producers of recorded sound (including performers and recording engineers), thus obscuring the wider range of human beings who were closely and quotidianly involved in the global phonographic network (including the anonymous women and children ruining their health in Indian shellac workshops, or the prematurely aged female workforce employed in record-pressing plants around the globe). Opening up the physical 'black boxes' of media cultures (as recommended by media archaeology), rather than a merely formal procedure, may therefore also enable us to retrieve what Marx described as 'the congealed forms of past knowledge and skills ossified in the form of machinery'.[45] The moment of disclosing may allow for a very partial posthumous reparation and acknowledgement to take place. For Cubitt, '[t]he dead are not remote, buried or lost: they are right under our hands, the concentrated dead labour in every tool and technology we handle'.[46] Yet, it is obvious that 'the voices of the dead cannot heal us by being ventriloquised: any attempt to let them speak is in fact a way of speaking on their behalf, so betraying the specificity of their

---

43   Anderson, Dunlop and Smith (2015), 4.
44   Parikka (2015), 37; 4.
45   As rephrased in Cubitt (2020), 30.
46   Ibid., 30–31.

lives'.[47] While the dead cannot be artificially re-called, something of their anonymous labour persistently resurface in artefacts so that the material present becomes both the medium and the site of a return. Whilst recognising unbridgeable differences between past and present technological realms, the transversal media-material approach thereby invites us to listen for effects of recurrence, interference, latency, and spectrality across temporal eras. In such an archaeological perspective, history is topographically understood as 'a *superimposition* of layers, which were successively sedimented but work in synchronous partnership'.[48] Archaeological regression may ultimately constitute the only means of accessing the present where the past reveals itself not as a repetition but as a beginning or a co-presence.[49] If the past becomes the medium of the present, the reciprocal is true. It follows that archaeology may liberate a past that 'has never been' or 'was never willed', releasing as it were its unlived futures.[50] Once revealed in the actuality of the present, the potentials of the past cannot be ignored, enclosed, or safely poured back into a fantasised historical bottle. Rather, they become urgently contemporary and must be counted with. Similarly, there is no going back after the black box has been opened.

The undisciplined, methodologically daring field of media archaeology has been key in fostering a range of new approaches to media technologies – and in particular to sound recording devices – in recent years. Many of these studies have been facilitated by the availability and digital (retro-) circulation of previously inaccessible archival material and the development of digital tools to unlock their contents. For instance, in 2008, the 'First Sounds' scientific team at the Lawrence Berkeley National Laboratory in California succeeded in visually recovering audio signals from one of Edouard-Léon Scott de Martinville's pre-Edison phonautograms, relying on 'painstaking optical imaging, conversion, and reconstruction'.[51] In the past decade, the study of phonography has increasingly been characterised by its commitment to interdisciplinary thinking and the pursuit of singular case studies as opposed to the 'projection of generalized theories'.[52] In these combinatory studies, the emphasis is – expectedly – placed on in-betweens and moments of encounters, co-presence and co-dependency. Recent studies have drawn attention to the material infrastructures which have made the

47 Ibid., 34.
48 Citton (2017), 216; my translation.
49 What Gumbrecht calls an 'origin of the present' (2013) and Agamben a 'source' (2009).
50 Agamben (2009), 103.
51 Altergott (2021), 20.
52 Ernst (2013), 44.

circulation of recorded sound possible, to processes of colonial extractivism, to the diverse sites, actors, objects and knowledge practices involved in the transnational development of early phonographic networks, or yet again to the environmental cost of recorded sound in the wake of the Anthropocene.[53] What they offer is not so much a radically alternative history of the early recording industry than a complementary interpretation – which is inevitably and openly coloured by contemporary concerns (through the detour of the past). These studies, because they frequently emphasise the 'deep time of the media', can undermine some deep-rooted assertions and implicit biases of media history.[54] In addition to proposing an extended time frame, a geographical rescaling – or decentring – is also important for it may unsettle what Shannon Mattern calls media archaeology's 'prevailing Western orientation, its occasional "orientalist" treatment of curious devices from other cultures and times, its mostly male bibliographies'.[55] Embracing a broader chronological and geographical scope therefore appears as a politically-motivated gesture, stemming from the desire to critically understand the transnational, unequal and slow geopolitical formation of media cultures – long before a global 'mediarchy' crystallised.[56]

In her inspiring media archaeology of urban mediation, Mattern productively opens up the field of media archaeological thinking by expanding both the geographical and temporal focus of her enquiry. Her patient exploration of ancient and modern networked cities allows her to reveal the spatially diffuse pre-history of the modern digital megalopolis. In the process, she encounters and examines an extended constellation of media such as mud, electricity, clay, print, concrete, and the human voice. Here, the understanding of what constitutes a medium is deliberately elastic – the medium is understood in its modest, most accommodating form of *mediating object*.[57] Similarly, my approach to the media-material cultures of shellac posits media and matter as mutually and concretely bound together, though not equivalent. It is no surprise that such sticky, malleable substances as clay and mud should be so central to Mattern's enquiry. Indeed, mediation may be theorised as a spatio-material process and quasi-fusional practice of linking, binding or assembling disparate elements together. It is worth noting here that India is an ancient 'clay culture' and clay artefacts, manufactured for

53   See Devine (2019a); Devine and Boudreault-Fournier (2021); Denning (2015); Radano and Olaniyan (2016); Roy (2021a); Silva (2016); Roy and Rodríguez (2021); Smith (2015).
54   See Zielinski (2006).
55   Mattern (2017), xxiii.
56   See Citton (2017).
57   Dant (1999), 154.

millennia, came to be regarded as both containing and celebrating 'nature's primordial energies and growth processes'.[58]

Just like clay, a material sometimes described as the first plastic,[59] shellac is a natural and ancient binding medium, present across the everyday cultures of India (particularly northern and central India). It is an adhesive or binding material which invisibly cements together disparate physical and symbolic elements. As such, it ambiguously wavers *between* materiality and mediality, thus occupying an intermediary category (that of adhesive) that media theory has yet to fully reckon with. Accordingly, the present study highlights what I call the 'adhesive' dimension of mediation, drawing attention to media cultures as concretely and dynamically bound together on a physical as well as symbolic level. In this context, it is worth keeping in mind media and literary theorist Samuel Weber's definition of medium as that which simultaneously binds and separates – both materially and symbolically: 'The medium is [...] distinguished on the one hand from a simple emptiness, on the other hand from the impenetrability of matter: it divides and connects at the same time, more precisely: *it only makes the connection possible as division*'.[60] I will return to these interdependent notions of separateness and connectedness throughout the book. We will see that shellac could serve, for instance, as a binding substance (in the case of gramophone records) or as an insulant (in the case of grenades, and munitions) – bringing disparate elements (and people) together or, on the contrary, keeping them hermetically separated from one another.

Though not a strict media archaeology of shellac, the present book offers a contribution to the expanding field of media-material theory. And though its subject matter is limited to one specific material, I hope that this study may be furthered and challenged by supplementary research into other materials of culture. Existing studies of phonographic materials notably include cultural analyses of polyvinyl chloride (PVC), of Carnauba wax, or yet again of mica – an incomplete and rapidly expanding list.[61] Social geographers Chris

58 Bussabarger and Robins (1968), 7.
59 See Bell (1936), xi. It must be noted that clay is to be found in abundance in the river valleys of the Ganges. On the invention of clay modelling, see Pliny's anecdote, as related in Dillard (1976 [1974]), 64.
60 Weber quoted in Ernst (2013), 105; emphasis is mine.
61 See Westermann (2013) and Devine (2019a) on PVC; Silvers (2018) on carnauba wax; Bronfman (2021) on mica. A 2021 research workshop entitled 'Sound Supplies: Raw Materials and the Political Economy of Instrument Building' (organised remotely by Fanny Gribenski, Viktoria Tkaczyk and David Pantalony at the IRCAM in Paris) further engaged participants to reflect on 'the neglected history of the supply chains that made modern music cultures and audio communication possible' (as stated in the workshop's programme). As part of the event, participants contributed

Gibson and Andrew Warren have further chronicled the material history of
the guitar 'by physically following the wood': their ethnographical journey
took them 'from guitar to factory, factory to sawmill, sawmill to forests, and
eventually to the trees'.[62] In the academic field, most notably within media
and music studies, the close examination and rehabilitation of materials – as
well as the acknowledgement of the violence inherent to their cultivation,
extraction and circulation – therefore seems to be well under way. A number
of recent studies – notably in musicology – have precisely taken shellac as a
case-study as part of larger attempts to excavate the material, ideological and
perhaps even mythical infrastructures underpinning the global circulation
of recorded sound in the early twentieth century, while artists themselves
have started reengaging with discarded materials of culture (see Chapter 5).[63]
Moreover, after decades of cultural latency and obsolescence, it must be
noted that commercial interest in shellac has been revived in recent years,
as part of industrial research into sustainable bioplastics: its renewable,
biodegradable, and non-toxic properties make it a valuable component in
the contemporary food, pharmaceutical and cosmetic industries.

Sources and transdisciplinary crossings

Media archaeologists, art historians and material culture theorists have
long engaged with the materiality of making, grappling as they do with
the 'cultural logics of materials' and their concrete, intricate messiness and
resistance.[64] This book draws from heterogeneous primary and secondary

networked histories of musical substances ranging from carbon black, ivory, paper, and steel
to rubber and wood. The year before, a symposium on latex and the logic of extractivism was
organised at ICI Berlin, examining the 'ruinous consequences' which rubber extraction had on
the Amazon Forest and its populations. The symposium was entitled 'Latex. Critical Inflections
on (Neo)Extractivism in Latin America'.

62   Gibson and Warren (2021), 5; 4.

63   On phonographic networks, see Smith (2015); Devine (2019a); Roy (2021); Williams (2021). In
2017, the Bauhaus Dessau Foundation held a design exhibition devoted to 'smart materials' – a
term typically associated with digital culture but which included, in a retro-futuristic twist, a
number of premodern materials including shellac, repositioning it as a material for the digital age
and inscribing it within contemporary debates on eco-materials. The exhibition, entitled *smart
materials satellites. Material als Experiment* (*smart materials satellites. Material as Experiment*),
notably displayed the works of German designer duo Lapatsch|Unger (see Chapter 5). Shellac was
also used in Berlin's Staatsoper opera house as part of a 2017 refurbishment aimed at improving
its acoustic properties. I am grateful to Karsten Lichau for pointing this out to me and excavating
an article on the topic published in the *Berliner Morgenpost*.

64   Anderson, Dunlop, Smith (2015), 12.

sources, in keeping with the belief that meaning may emerge dialogically from the encounter and confrontation of a variety of texts and objects (understood in the broadest possible sense). Beside archival material, a wide range of scholarly texts including those of cultural history, media theory, music studies and art history have been consulted. The transdisciplinary lines of enquiry which are developed throughout the book – and across each individual chapter – have stemmed from the material itself: taking Tim Ingold's invitation to 'follow the materials' of culture as a heuristic prompt, I have been 'following' shellac in its several guises and shapes – interrogating objects and practices as well as the stories which are told about them –, so as to attend to its always emerging material history.[65] This produces a methodically 'unsettled' or dynamic study that 'remain[s] ever alert to visual and other sensory cues in an ever changing environment'.[66] The book therefore makes room for accidents and chance encounters: to some extent, it is also partially produced by such encounters.

Ingold's proposal to follow the materials resonates with psychosocial theorist Lisa Baraitser's reflection on 'the kind of freedom of movement that allows untethered concepts, texts, ideas, objects, practices or methods to cross disciplinary domains' – yet such a movement, she notes, is never fully untethered or historically autonomous.[67] Baraitser's ambition is not to liquidate or dismiss disciplinary traditions but to tentatively reveal (or liberate) the epistemic potentials and values to be found in the (experiential) moment of 'discovery' itself, before experience becomes sedimented and classified. For Baraitser, the point of interdisciplinary practices is not to provide definitive statements, close systems, replicable models or syntheses but rather to realise the open-ended, precarious and emergent quality of thinking. While not fully operating outside disciplinary fields, an object-driven methodological reorientation or recalibration may therefore lead to the formation of thought-provoking disciplinary hybrids. Baraitser compares the 'trans-' of transdisciplinary studies to what is known, in chemistry, as a 'free radical':

Here an atom has an open electronic shell, making free radicals chemically promiscuous with others, and also with themselves, highly reactive, transformational. The bonds are suggestively described by chemists as 'dangling', somehow available for polymerisation as they move. So, as a

---

65   Ingold's approach expands upon geographer Ian Cook's recommendation to 'follow' everyday objects (2004).

66   Ingold (2010), 94.

67   Baraitser (2017), 30.

concept departs from one disciplinary domain and inserts itself in another, it may both underscore the distinction between those domains, whilst at the same time, through its anomalous presence, bring about some kind of change or re-formation.[68]

Guided by Baraitser and Ingold, I am therefore interested in asking what a methodological shift of emphasis – from the study of well-defined and familiar cultural objects to the study of materials themselves – might allow us to 'do' (and undo) from a theoretical viewpoint, across material culture studies and media theory. More than a linguistic game, the apparently simple displacement from 'material culture' to 'materials of culture' therefore yields larger and deeper implications for the understanding of past and contemporary media realms.

Chapter overviews

This book embraces some of the unstable stories, histories and materialities of shellac. It focuses on various moments in turn, moving through various shapes and mediatic uses of the material, across the visual and the sonic realm. Every chapter provides a different perspective on one single (yet mutable and polysemic) substance which came to bear different meanings and values in different contexts. Every time, the particular emphasis contributes to shedding light on one broad techno-cultural era, while drawing attention to the deeper relation of interdependence – or synchronicity – which may exist between very different moments in time and space. As such, no rigid distinction is posited between the premodern and the modern, but the latter is understood – in Hartmut Böhme's terms – as 'embody[ing] the presence of all previous historical periods'.[69] Reciprocally, it may be that 'modernity can only begin to understand itself if it makes use of cognitive resources from epochs considered to be premodern':[70] understanding the present moment may only be (partially) possible through a movement of recursion.

The chapters follow a roughly chronological order, describing five specific (and interdependent) moments in the long history of the material. While the sites I describe function together and sometimes overlap, they can also be approached as self-contained, but necessarily incomplete, fragments. The point

68 Ibid.
69 Böhme (2014), 14.
70 Ibid.

here is not to create fractions or divisions, rather to engage with the episodic, discrete, cumulative formation of what cannot – and should not – be seen as a unified narrative or as a deceptively coherent aggregate.[71] Accordingly, each chapter functions as a temporary and tentative container, which also contributes to shaping the material. These shapes, however, are not mutually exclusive, definitive or 'finished' but mutually and cumulatively transform one another.

Long before Emile Berliner actualised its potentials for sound reproduction in the mid-1890s, shellac (as well as lac dye) played a significant role in a number of cultural visual practices in South Asia and in Europe. Chapter 1 investigates the early uses, cultural understandings and traditional applications of lac and shellac as visual media in India (where it was notably used to decorate the body in practices of self-inscription). It retraces how the resources were imported by Dutch merchants and the British East India Company in the early 1600s, paying attention to the geopolitical infrastructure which authorised their circulation. The chapter describes how shellac was speculatively translated, transformed and reinterpreted in the European context. In particular, attention is paid to the reflective and imitative properties of the medium – and to the intuition of its sonic properties (when Italian violin makers began using it to varnish musical instruments). The chapter shows how the devaluation of lac dyes in the mid-nineteenth century led to the reconsideration of shellac, paying attention to the discovery and commercial exploitation of its plasticity in the second half of the nineteenth century.

Chapter 2 retraces how shellac progressively and predominantly became a medium of sound in the late nineteenth century, focusing on Emile Berliner's discovery of its sonic properties. The chapter especially focuses on the US where the resource was progressively domesticated and 'Americanised', to the point of erasing its provenance and pre-mediatic histories. Yet I suggest that the novel media artefact of the disc remediated some of the earlier shapes of shellac, emphasising in particular its material and symbolic affinities with seals, masks and statues, and discussing the relation between antiquity and modernity (notably in connection with Emile Berliner and Jean Cocteau, who both highlighted the correspondences between sound reproduction and previous techniques of memorialisation). The chapter also shows that the culturally standardised gramophone record was not materially standardised, stressing how ideals of sonic perfection were predicated upon the imperfect grain of the record. An important aspect of this chapter is that it makes visible the forms of labour entombed in the

---

71 Beyond the critique of 'flat ontology'.

commodity of the record. It offers a parallel between shellac production in Indian workshops and the work carried out in western pressing plants, notably insisting on the crucial contribution of female labourers in the early phonographic industry. Mapping out the relationship between shellac workshops and gramophone factories – and their belonging to the same ideological continuum – allows me to partially expose the implicit colonial infrastructure underpinning the early recording industry.

While Chapter 2 insists on the sonicity of shellac, Chapter 3 surveys the intersensory position of recorded sound in the interwar period – also conceived of as a 'golden age' of shellac. It notably does so through recovering the largely forgotten – yet significant – trope of the 'mirror of the voice' and surveying how it was materially and discursively interpreted by groups as diverse as theorists, artists and home recordists. The first section of this chapter discusses the visual phono-fetishism of the interwar period and critically reengages with Adorno's essays on phonography (where he notably explored the relation between identity, recorded sound and self-alienation). As a counterpoint, the second part of the chapter attends to the defetishising discourse offered by interwar art and design practices (notably those carried out at the Bauhaus in Weimar), exploring how they contributed to creating a new, intermedial understanding of phonography – between sound and vision. The chapter also proposes that a theoretical shift from the hauntological (or spectral) to the specular – and from reproducibility to reflectivity – may yield important insights for the understanding of interwar phonographic cultures.

Chapter 4 attends to the toxic transformation of shellac in the two World Wars, when it became a key substance in the manufacture of detonating compositions, hand grenades, and bombs and was rationed by Western governments (thus curtailing record-making operations and intensifying research into substitutes, including PVC). Drawing from Malabou's radical theses on 'destructive plasticity', it theorises the material and ideological instability of shellac as well as its recycling, exploring the dominant discourses associated with recorded sound.[72] In particular, it draws attention to the trope of phonographic listening as a means to repair both individual and social bodies broken down by war, showing how this discourse was recuperated by governmental bodies. Parallel phonographic practices – such as the recording sessions which took place in German prisoner-of-war camps (such as the 'Halfmoon Camp' in Wünsdorf) – are also discussed.

Chapter 5 explores the recurrence and persistence of shellac in contemporary art and design, describing its visual and material remediation

---

72 See Malabou (2012).

by contemporary practitioners based in Germany, France, Britain and the US. Their creative works are notably discussed in relation to the *Broken Music* exhibition (1988–1989), first shown in Berlin, which marked a critical turning point in what could be called 'gramophone art'. Throughout, the chapter discusses the importance of embodied modes of knowing for the exploration of materiality – and revives Dagognet's invigorating plea for an ontology of (neglected) materials.

Bibliography

Acland, Charles R., ed. 2007. *Residual Media*. Minneapolis and London: University of Minnesota Press.

Adarkar, Bhalchandra P. 1945. *Report on Labour Conditions in the Shellac Industry*. Delhi: Indian Labour Investigation Committee.

Agamben, Giorgio. 2009. *The Signature of All Things: On Method.* Translated by Luca D'Isanto with Kevin Attell. New York: Zone Books.

Altergott, Renée. 2021. 'Une machine à gloire? Legacies of the French Inventor(s) of Sound Recording through the Ages'. *French Forum* 46 (1): 19–35.

Anderson, Christy, Anne Dunlop and Pamela H. Smith, eds. 2015. *The Matter of Art: Materials, Practices, Cultural Logics, c. 1250–1750*. Manchester: Manchester University Press.

Asendorf, Christoph. 1993. *Batteries of Life: On the History of Things and Their Perception in Modernity.* Translated by Don Reneau. Berkeley, Los Angeles, London: University of California Press.

Barad, Karen. 2017. *Meeting the Universe Halfway: Quantum Physics and the Entanglement of Matter and Meaning*. Durham and London: Duke University Press.

Baraitser, Lisa. 2017. *Enduring Time*. London, New York, Oxford, New Delhi, Sydney: Bloomsbury Academic.

Barry, Andrew. 2005. 'Pharmaceutical Matters: The Invention of Informed Materials'. *Theory, Culture & Society* 22 (1): 51–69.

Bell, L. M. T. 1936. *The Making & Moulding of Plastics.* London: Hutchinson's Scientific & Technical Publications.

Benjamin, Walter. 1973. *Illuminations.* London: Fontana.

Bennett, Jane. 2010. *Vibrant Matter: A Political Ecology of Things.* Durham and London: Duke University Press.

Böhme, Hartmut. 2014. *Fetishism and Culture: A Different Theory of Modernity.* Translated by Anna Galt. Berlin and Boston: De Gruyter.

Bolter, David Jay and Richard Grusin. 1999. *Remediation: Understanding New Media.* Cambridge, Massachusetts: The MIT Press.

Bronfman, Alejandra. 2021. 'Glittery: Unearthed Histories of Music, Mica, and Work'. In *Audible Infrastructures: Music, Sound, Media,* eds. Kyle Devine and Alexandrine Boudreault-Fournier, 73–90. New York and Oxford: Oxford University Press.

Bussabarger, Robert F. and Betty D. Robins. 1968. *The Everyday Art of India.* New York: Dover Publications.

Citton, Yves. *Médiarchie.* 2017. Paris: Editions du Seuil.

Cook, Ian. 2004. 'Follow the Thing: Papaya'. *Antipode* 36: 242.

Crockett, Clayton. 2010. 'Foreword', *Plasticity at the Dusk of Writing: Dialectic, Destruction, Deconstruction,* by Catherine Malabou. Translated by Carolyn Shread, xi–xxv. New York: Columbia University Press.

Cubitt, Sean. 2013. 'Anecdotal Evidence'. *NECSUS_ European Journal of Media Studies.* Available from https://necsus-ejms.org/anecdotal-evidence/.

Cubitt, Sean. 2017. *Finite Media.* Durham: Duke University Press.

Cubitt, Sean. 2020. *Anecdotal Evidence: Ecocritique from Hollywood to the Mass Image.* New York: Oxford University Press.

Dagognet, François. 1997. *Des détritus, des déchets, de l'abject: Une philosophie écologique.* Le Pressis-Robinson: Institut Synthélabo.

Dant, Tim. 1999. *Material Culture in the Social World: Values, Activities, Lifestyles.* Maidenhead: Oxford University Press.

Daston, Lorraine. 2004. 'Introduction: Speechless'. In *Things That Talk: Object Lessons from Art and Science,* ed. Lorraine Daston, 9–24. New York: Zone books.

Denning, Michael. 2015. *Noise Uprising: The Audiopolitics of a World Musical Revolution.* London, New York: Verso.

Derrida, Jacques. 1994. *Spectres of Marx.* New York and London: Routledge.

Devine, Kyle. 2019a. *Decomposed: The Political Ecology of Music.* Cambridge, Massachusetts and London, England: The MIT Press.

Devine, Kyle and Alexandrine Boudreault-Fournier, eds. 2021. *Audible Infrastructures: Music, Sound, Media.* New York and Oxford: Oxford University Press.

Dillard, Annie. 1976 (1974). *Pilgrim at Tinker Creek.* London: Pan Books.

Domínguez Rubio, Fernando. 2016. 'On the Discrepancy between Objects and Things: An Ecological Approach'. *Journal of Material Culture* 21(1): 59–86.

Ernst, Wolfgang. 2011. 'Media Archaeography: Method and Machine versus History and Narrative of Media'. In *Media Archaeology: Approaches, Applications, and Implications,* eds. Erkki Huhtamo and Jussi Parikka, 239–255. Berkeley: University of California Press.

Ernst, Wolfgang, 2013. *Digital Memory and the Archive.* Minneapolis and London: University of Minnesota Press.

Ernst, Wolfgang. 2016. *Sonic Time Machines: Explicit Sound, Sirenic Voices, and Implicit Sonicity.* Amsterdam: Amsterdam University Press.

Fuller, Matthew. 2005. *Media Ecologies: Materialist Energies in Art and Technoculture.* Cambridge, Massachusetts: The MIT Press.

Gettens, Rutherford J. and George L. Stout. 1966 [1943]. *Painting Materials: A Short Encyclopaedia.* New York: Dover Publications.

Gibson, Chris and Andrew Warren. 2021. *The Guitar: Tracing the Grain Back to the Tree.* Chicago and London: The University of Chicago Press.

Gumbrecht, Hans Ulrich. 2013. *After 1945: Latency as Origin of the Present.* Stanford, California: Stanford University Press.

Herzogenrath, Bernd. 2015. 'Media|Matter: An Introduction'. In *Media|Matter: The Materiality of Media|Matter as Medium*, ed. Bernd Herzogenrath, 1–16. New York: Bloomsbury Academic.

Hockenberry, Matthew, Nicole Starosielski and Susan Zieger, eds. 2021. *Assembly Codes: The Logistics of Media.* Durham and London: Durham University Press.

Ingold, Tim. 2010. 'The Textility of Making'. *Cambridge Journal of Economics* 34: 91–102.

Ingold, Tim. 2012. 'Towards an Ecology of Materials'. *The Annual Review of Anthropology* 41: 427–442.

Jones, Geoffrey. 1985. 'The Gramophone Company: An Anglo-American Multinational, 1898–1931'. *The Business History Review* 59 (1): 76–100.

Kouwenhoven, John A. 1982. 'American Culture: Words or Things?'. In *Material Culture Studies in America*, ed. Thomas J. Schlereth, 79–92. Nashville: American Association for State and Local History Press.

Kubler, George. 1962. *The Shape of Time: Remarks on the History of Things.* New Haven and London: Yale University Press.

Lehmann, Ann-Sophie. 2015. 'The Matter of the Medium: Some Tools for an Art-Theoretical Interpretation of Materials'. In *The Matter of Art: Materials, Practices, Cultural Logics, c. 1250–1750*, eds. Christy Anderson, Anne Dunlop and Pamela H. Smith, 21–41. Manchester: Manchester University Press.

Leslie, Esther. 2020. 'Devices and the Designs on Us: Of Dust and Gadgets'. *West 86th: A Journal of Decorative Arts, Design History, and Material Culture* 27 (1): 3–21.

Malabou, Catherine. 2010. *Plasticity at the Dusk of Writing: Dialectic, Destruction, Deconstruction.* Translated by Carolyn Shread. New York: Columbia University Press.

Marks, Laura U. 2000. *The Skin of the Film: Intercultural Cinema, Embodiment and the Senses.* Durham and London: Duke University Press.

Mattern, Shannon. 2017. *Code and Clay, Data and Dirt: Five Thousand Years of Urban Media.* Minneapolis and London: University of Minnesota Press.

Parikka, Jussi. 2010. *Insect Media: An Archaeology of Animals and Technology.* Minneapolis and London: University of Minnesota Press.

Parikka, Jussi. 2015. *A Geology of Media.* Minneapolis and London: University of Minnesota Press.

Perec, Georges. 1989. *L'infra-ordinaire*. Paris: Seuil.

Petroski, Henry. 1989. *The Pencil: A History of Design and Circumstance*. London and Boston: Faber & Faber.

Radano, Ronald and Tejumola Olaniyan, eds. 2016. *Audible Empire: Music, Global Politics, Critiques*. Durham and London: Duke University Press.

Roy, Elodie A. 2021a. 'Another Side of Shellac: Cultural and Natural Cycles of the Gramophone Disc'. In *Audible Infrastructures: Music, Sound, Media*, eds. Kyle Devine and Alexandrine Boudreault-Fournier, 207–226. New York and Oxford: Oxford University Press.

Roy Elodie A. and Eva Moreda Rodríguez, eds. 2021. *Phonographic Encounters: Mapping Transnational Cultures of Sound, 1890–1945*. Oxon and New York: Routledge.

Schwenger, Peter. 2006. *The Tears of Things: Melancholy and Physical Objects*. Minneapolis: University of Minnesota Press.

Silva, João. 2016. *Entertaining Lisbon: Music, Theatre, and Modern Life in the Late 19th Century*. New York: Oxford University Press.

Silvers, Michael. 2018. *Voices of Drought: The Politics of Music and Environment in Northeastern Brazil*. Urbana, Illinois: University of Illinois Press.

Schmidgen, Henning. 2022. *Horn, or the Counterside of Media*. Durham and London: Duke University Press.

Smith, Jacob. 2015. *Eco-Sonic Media*. Oakland, California: University of California Press.

Sobchak, Vivian. 2004. *Carnal Thoughts: Embodiment and Moving Image Culture*. Berkeley, Los Angeles, London: University of California Press.

Spate, Oskar H. K. 1964 (1954). *India & Pakistan: A General and Regional Geography*. London: Methuen.

Steinbeck, John. 1997 (1962). *Travels with Charley*. London: Arrow Books.

Stewart, Susan. 2007 (1993). *On Longing: Narratives of the Miniature, the Gigantic, the Souvenir, the Collection*. Durham and London: Duke University Press.

Thompson, Michael. 1979. *Rubbish Theory: The Creation and Destruction of Value*. Oxford: Oxford University Press.

Westermann, Andrea. 2013. 'The Material Politics of Vinyl: How the State, Industry and Citizens Created and Transformed West Germany's Consumer Democracy'. In *Accumulation: The Material Politics of Plastic*, eds. Jennifer Gabrys et al., 68–86. London and New York: Routledge.

Williams, Gavin. 2021. 'Shellac as Musical Plastic'. *Journal of the American Musicological Society* 74 (3): 463–500.

Zielinski, Siegfried. 2006. *Deep Time of the Media*. Translated by Gloria Custance. Cambridge, Massachusetts and London, England: The MIT Press.

1. Sheen: Early stories and circulation of shellac

Abstract:

Chapter 1 investigates the early uses, cultural understandings and traditional applications of lac and shellac as visual media in India (where it was notably used to decorate the body in practices of self-inscription). It retraces how the materials were imported by Dutch merchants and the British East India Company in the early 1600s, paying attention to the geopolitical infrastructure which authorised its circulation. The chapter describes how the medium was speculatively translated, transformed and reinterpreted in the European context. In particular, attention is paid to the reflective and imitative properties of the medium – and to the intuition of its sonic properties (when Italian violin makers began using it to varnish musical instruments).

Keywords: colonialism, India, shellac, East India Company, visual culture

The story of shellac begins on a local, infinitely small scale. The resource which, once processed, became a key component in global record manufacturing at the turn of the twentieth century was secreted by a tiny parasitic insect. The *Tachardia lacca* or *Laccifer lacca*, as the insect was scientifically described in the eighteenth century, measures no more than 1/20th of an inch long.[1] It infests a variety of host trees – principally kusum, palas and ber – most commonly found in the forest region lying between Calcutta and Central India, and yields two principal by-products.[2] The first one of these is a reddish dye, known as lac. The second one, shellac, is an orange-coloured resinous substance exuded by the clustered bodies of breeding females

---

1   Knaggs (1947), 208.
2   Melillo (2014), 239.

Roy, E.A., *Shellac in Visual and Sonic Culture: Unsettled Matter.* Amsterdam: Amsterdam University Press, 2023

DOI 10.5117/9789463729543_CH01

infesting the fresh twigs of the host tree.[3] In the course of their six-month existence, the blind and wingless female insects, settling 'in colonies of thousands'[4] on the new growth of trees, build an intricate architecture of hard resinous cases and passageways. This labyrinthine infrastructure is the place where the insects live and give birth, but it also serves as their graveyard.[5] The new generation, unable to settle on the parent tree, swarm and find other trees to infest.[6] The cycle begins again. The lac-encrusted branches of the old tree are manually harvested and form the basis for shellac production (see Chapter 2). The production of shellac – and by extension the history of sound reproduction – is closely linked to the cycles of productivity and fertility of the insects. In a 1947 book devoted to the disappearing age of natural plastics and waxes, Nelson S. Knaggs poetically observed how lac bugs 'make their own coffins or mummy cases, which not only serve as tombs for their dead bodies and act as fortresses against their enemies, but also as incubators for the next generation of insects'.[7] Here, these two poles, emergence and eradication, seem to coincide. In later parts of this book, I argue that there are parallels to be drawn between the buried (yet life-imparting) insects, and the technologically disembodied voices – and forms of labour – entombed within the grooves of the gramophone disc.

This chapter recovers the ways in which lac and shellac – the two main by-products of the female lac bug – were routinely utilised in the South Asian and European contexts prior to the twentieth century. In doing so, it reveals some of the early histories and imaginaries of shellac that were subsequently repressed and deactivated. Lac was traditionally valued for its colouring properties (it yielded a dark red, sanguine hue), while the resin was prized for its adhesive powers, its shine and its plasticity. Indeed, shellac was understood and used as a plastic in South Asia long before interest in – and knowledge of – plastic materials began developing amongst western industrialists in the second half of the nineteenth century.[8] The chapter describes a pre-mediatic (and pre-sonic) moment in the long history of the material, revealing a series of early cultural and epistemic practices

---

3   See Melillo (2020).
4   Baxter (1954), 128.
5   Berenbaum (1995), 120–121.
6   Baxter (1954), 128.
7   Knaggs (1947), 208.
8   In the twentieth century, hundreds of shattered shellac ornaments (including bracelets and beads) spanning a period 'from the 3rd century B. C. to the end of the Pala period' were unearthed from the archaeological site of Chandraketugarh (situated in the north-east of Calcutta), now preserved in Calcutta's Ashutosh Museum. See Sinha (1966), 4.

which – though effectively distant and irreducible to it – can nevertheless help us understand some of the conditions of emergence of the gramophone disc. It highlights the concrete, embodied formation of scientific knowledge, stressing in particular its circulatory and participatory aspects. As persuasively proposed by Kapil Raj, a 'focus on circulation' – understood as a 'site of knowledge formation'– allows to address 'the *mutable* nature of the materials – of the men themselves and of the knowledges and skills which they embodied – as also their transformations and reconfigurations in the course of their geographical and/or social displacements'.[9]

Though they became progressively forgotten and discarded in the contemporary era, the pre-mediatic stories of shellac still had currency in late nineteenth-century culture. For instance, Emile Berliner, as he envisioned the shellac disc as 'the seal of the human voice', was explicitly reigniting a well-known association between shellac, inscriptibility and memory – as well as taking full advantage of the modern technique of industrial thermocasting developed around 1850.[10] Although the record as technical object involved infinitely more than a linear or anecdotal 'remediating' of the seal, there exist material and symbolic continuities – and contiguities – between pre-mediatic shellac artefacts and gramophone discs. The instance of recycling, however, cannot be limited to the punctual recuperation or redeployment of discrete 'shapes' or objects. On a larger, infrastructural level, it may be proposed that some of the wider colonial routes and discourses which allowed for materials and goods to be circulated across Europe (and North America) in the first place residually informed and authorised the transnational phonographic network of the early twentieth century. The example of shellac unambiguously demonstrates how '[m]edia history participates in stories of global expansion through colonialism and the rush for resources'.[11] The continuous material and epistemic exchanges between Europe (especially Great Britain) and India, and their long economic and political entanglement up until the aftermath of the Second World War, further invite us to consider how the proto-mediatic or early visual regime of shellac continued to stir beneath its sonic becoming – in a manner that is best described as 'parasitic'. Accordingly, this chapter is predominantly concerned with the itinerant stories, structures of knowledge and infrastructures which shaped early cultures of shellac. While it acknowledges the historical and geographical specificity of these early narratives and epistemic networks,

9   Raj (2007), 20–21; emphasis in the original text.
10 See Hughbanks (1945), 19.
11 Parikka (2015), 26.

I also show how some of these would endure well into the contemporary era (albeit in a diluted or implicit, rather than explicit, form). This first chapter therefore serves as a matrix: it introduces a number of thematic patterns – such as the connection between shellac and femininity, and shellac and mnemonic inscription – which will be reencountered and discussed in the rest of the book.

A hundred thousand stories: Early narratives and poetic accounts

The early stories of lac and shellac – as recounted by Sanskrit poets and thinkers including Kalidasa, Bharavi and Panini – reveal in particular their enduring associations with the realms of fertility and femininity,[12] emphasising shellac's mythopoetic connection to forms of abundance or excess (a point worth keeping in mind if we recast the record industry as a history of surplus, accumulation, and waste).[13] The Sanskrit term for the insect, *lakšā* – meaning 'a hundred thousand' – refers to the fertile proliferation of bugs infesting tree branches. The plentiful *lakšā* is to be linked to the sacred figure of *Lakšmi* – both the Hindu goddess of prosperity and beauty, and the wife of *Višnu*, symbolising the principle of duration and eternity.[14] The lac insect appears as the embodiment of *Lakšmi* on earth or, as a Vedic prayer praising the female lac insect put it, 'as a beautiful young maiden'.[15] Art historians Bussabarger and Robins, surveying the techniques and everyday materials (including clay, fibre and wood) of Indian arts and crafts, have emphasised their simultaneously sacred and secular character, demonstrating how all of the earth's inorganic and organic materials were 'involved in the Hindu concepts of existence and religion' so that 'natural phenomena evoked spiritual interpretations'.[16]

The *lakšā* insects secreted substances which entertained, from the very start, an intimate and symbiotic relationship to the human body – and particularly the female body – as well as to practices of repair and self-inscription. Several medicinal treatises from the Vedic period offer recipes for shellac-based medicinal oils and ointments, describing the host plant and the benefits of lac as medicinal potion and wound healer.[17] As anthropologist

---

12   See Sinha (1966).
13   On recorded sound and the logic of waste, see Roy (2020).
14   Bussabarger and Robins (1968), 197.
15   Mandal (n.d.), 1–2.
16   Bussabarger and Robins (1968), 119.
17   Mandal (n.d.), 1–2.

Sukumar Sinha gathered from his extensive bibliographical research into the early practices and imaginaries of the material, the easily digestible lac was believed to help cure ailments as diverse as 'cough, biles, polluted blood, hiccup, consumption, fever, pimples, erysipelas and leprosy'.[18] It was believed that it could heal and strengthen failing organisms, or cosmetically enhance them. While shellac could be taken internally (literally in-corporated) or externally displayed (in the form of bodily ornaments such as bangles[19]), red lac was superficially applied to the skin. Hindu women would soak a small, fibrous cotton pad (*alta*) with the red dye to ritually colour the palms of their hands, and the soles of their feet.[20]

The Sanskrit poem *Çiçoupâla-badha*, written by Māgha in the seventh century and first translated by the French philologist Hippolyte Fauche in 1863 (at the peak of Orientalism), contains several allusions to this cosmetic and gendered use of the material.[21] The poet keeps returning to the same image: a young woman is pictured walking away, barefoot, 'dyeing the earth with lac'.[22] Her lover gazes after the trail of lac-coloured footsteps – recording both her presence and her disappearance. The red footsteps act as precarious memory traces or literal land-marks, producing an ephemeral form of spatial inscription – a particular topography of love. They are reminiscent of the more permanent 'negative handprints' found by palaeontologists and archaeologists on the walls of prehistoric caves of Europe and South East Asia. Both the footprints and the handprints are bodily inscriptions or self-inscriptions, constituting an anonymous – yet poignant – repertory of pictorial mnemonic traces. In Màghat's poetic cycle, the recurrent red footprints function as nonverbal cues – creating a visual text within or beneath the written text, especially as the voice of the female protagonist is repressed or non-existing (she does not speak or write). It is worth recalling here that for folklore historians Robert Bussabarger and Betty Robins, the act of painting (including self-painting) constitutes a language in its most elemental form – a point of contact, no matter how transient and precarious, with alterity. In their survey of Indian pictorial folk traditions, they propose that '[b]y painting the human body, domesticated animals, stones, leaves,

18   Sinha (1966), 5.
19   Shellac ornaments were notably used as wedding gifts to signify a transition from maidenhood to womanhood. The bride would traditionally receive a pair of fragile *Ruli* bangles from her father as a wedding present – the bangles were worn to indicate her married status. See Sinha (1966), 36.
20   Mandal (n.d.), 3. The dye was also used to hand-print designs on calicos and prayer cloths.
21   The *Çiçoupâla-badha* was retitled the *Tétrade* by Fauche.
22   Fauche (1863), 204; my translation.

wood and the ground with organic stains, carbon and earth colors, men developed ways to diagram and eventually communicate their thoughts'.[23] As such, rather than being secret or private, acts of auto-inscriptions express (and may reciprocally originate from) the longing for another being to see, decipher and ultimately understand them. Decorating and altering one's body may reciprocally be construed as an opening of oneself to alterity, even when the other only exists as a virtual and hoped-for (rather than actual) presence.[24]

Beyond its highly symbolic cosmetic functions, more mundane applications of the material can be retraced. For instance, the voluminous *Ain-i-Akbari*, an official treaty providing detailed information on the governance of the Emperor Akbar's empire, gave precise instructions for shellacking the woodwork of public buildings.[25] The *Ain-i-Akari*, penned by court historian Abul Fazl and issued in 1590, was almost exactly contemporaneous with one of the earliest known accounts of shellac in Europe, published at the close of the century by Dutch merchant and historian Jan Huygen van Linschoten. Linschoten, who had travelled to India on a scientific mission initiated by the King of Portugal, compiled in his diary his observations on a number of indigenous craft practices. The account he published on his return bears a detailed description of the widespread process of applying shellac to wooden furniture and household objects – from bedsteads, chairs and desks to boxes – to protect them. The shellac resin could be used as a medium: once mixed with a pigment, it 'could be applied to in [its] hot, molten state to the surface to be coloured'.[26] Linschoten marvelled upon the 'beauty and brightness' of shellacked surfaces, comparing their smooth lustre to that of 'glass, most pleasant to behold'.[27] The analogical rapprochement with glass was especially evocative in the context of late sixteenth century Dutch culture, at a time when the technique of using glass in the manufacture of mirrors – famously perfected in the 1560s by a group of Venetian mirror-makers – was just being taken up by craftsmen in the Netherlands.[28] Through

23   Bussabarger and Robins (1968), 146.
24   This could be related to Michel Serres's discussion of cosmetics in relation to Pierre Bon-nard's 1931 painting *Nu au miroir*. Serres proposes that the practice of making up is a means of making the 'the real skin become visible, as it is lived within oneself' (Serres [1985], 31; my translation) – rather than a strategy of concealment or repression. There is no division for Serres between surface and depth (a reading which contrasts with the dismissal of cosmetics as a sign of duplicity and falseness in Plato's *Gorgias*).
25   Parry (1935), 2.
26   Osborne (1975), 233.
27   Linschoten quoted in Parry (1935), 3; modernised spelling is mine.
28   Osborne (1975), 570–571.

Linschoten's analogy, the ancient medium thus became equated as it were with a strictly contemporary surface of modernity and imbued with its associations. Interestingly, a more tangible association between shellac and mirror-making would be made three centuries later, when a layer of shellac routinely served to (invisibly) '[attach] silver to the backs of mirror' whilst hand-held mirrors would often bear finely engraved frames or backs made of shellac (see Figure 2 below).[29] It was further believed that a shellac solution, finely sprayed on the silvered surface of the mirror, would 'add life' to it.[30]

But another association between shellac and (black) mirrors – of which Linschoten was apparently not aware – is to be found in the context of divinatory practices. Travel diaries and observations gathered by colonial travellers between the seventeenth and nineteenth centuries (including French official Louis Auguste Bellanger de L'Espinay and British Colonel Stephen Fraser) describe occult ceremonial rites involving shellac. The substance was poured into an earthenware vase and would be stirred by young girls over a fire, progressively transforming into a 'black liquid, viscous as the tar collected from volcanic cracks in the ground'[31] as it was slowly heated. The warm shellac film would subsequently be spread on the ground where, drying and solidifying, it formed a mirror-like surface in which one would reputedly be able to see once again the faces of loved ones – or grasp the shape of future events. For many ancient cultures, mirrors (or reflective surfaces) were actual as well as deeply spiritual – or even magical – artefacts. Their distinctive ritual significance was evidenced by a number of folktales and myths, including that of Narcissus and Echo (which is revisited in Chapter 3).[32] This secret, 'divinatory' dimension of the shellac medium – giving access both to the past and the future – is worth keeping in mind as it would irrepressibly surface again in the early imaginaries of phonography. Complementarily, connections between the realm of femininity, phonography and the occult were repeatedly drawn in the realm of popular culture where recorded sound was reimagined as a 'sympathetic medium' – the fictional gendering of the phonograph as female in late Victorian novels such as Villiers de L'Isle Adam's *L'Eve future* (1886) and George Du Maurier's *Trilby* (1894) is worth remarking here.[33] Furthermore, European visual artists of the interwar period repurposed the

---

29 Parry (1935), 173. See also Katz (1985), 23.
30 Hicks (1961), 227.
31 Maillet (2009), 58.
32 See Osborne (1975), 569; Melchior-Bonnet (1994), 114.
33 Galvan (2010), 99; see also Miller Frank (1995).

material as well as the discursive idea of the disc as a mirror (existing both
as a surface of reflection and mediation), while alternative psychoanalytical
models of subject formation offered, in answer to Lacan's ocular-centrism,
the hypothesis of an acoustic mirror stage (see Chapter 3).

European circulation and speculation

India had been trading with neighbouring countries, most notably China,
for centuries before the arrival of European merchants and shellac was
well-known within the South Asian world.[34] Although medieval dye recipes
suggest that lac was imported in Europe by the Catalans and Provençals
as early as 1220, the circulation of Indian goods in Europe only became
systematised with the establishment of the Portuguese's colony of Goa in the
early sixteenth century, and the foundation of the East India Companies at
the start of the following century.[35] Emboldened by the Portuguese success
and foreseeing flourishing trading opportunities – as well as wishing to
emancipate themselves from a reliance on Portugal –, all major European
powers aspired to secure a foothold in South Asia and thus strengthen
their positions on the world's market. Communities of English, Dutch,
and Danish merchants rapidly set out to establish their own competitive
East India Companies in the first two decades of the seventeenth century
(respectively in 1600, 1602, and 1616), during the time of the Mogul Empire.
The establishment of the EICs made the Indian subcontinent 'the pivot
of Asian maritime trade', an area rife with political tensions, revolts, and
commercial intrigues.[36] By the late seventeenth century, the principal
settlements were Goa (Portuguese), Calcutta (British), Chinsurah (Dutch),
all of those being strategically situated in the district of the Hooghly River
which provided access to the Indian Ocean. The French initially settled in
Surat (on the west coast) before moving eastwards to Pondicherry in 1671
and establishing themselves in Chandernagore in Bengal fifteen years later,
close to the English and Dutch settlements.[37]

---

34   Osborne (1975), 234.
35   See Eastaugh, Walsh, Chaplin and Siddall (2004), 214.
36   Raj (2007), 34.
37   See Raj (2007), 34–35. A relative latecomer, the French *Compagnie des Indes Orientales* was
officially founded in 1664 by a royal edict, differing in this respect from the other East India
Companies which had all been instigated by merchant communities. The aim of the *Compagnie
des Indes Orientales* was to secure, primarily for the French domestic market, commodities
which had previously been bought from English and Dutch merchants.

Shellac was first shipped from Indian ports – and particularly Calcutta, the epicentre of the British trade – to London, travelling on merchant vessels also transporting other valuable commodities such as tea, chintzes, silk, tin, spices, to name but a few.[38] The Company began building its own ships in 1607 in Deptford to lessen building costs, setting out to build larger and larger ships.[39] By 1621, it ran the biggest private fleet in England (totalling over twenty ships) and employed 2,500 seamen.[40] Before setting off on their long and hazardous journey (ships were not infrequently lost to pirates, to rival trading companies or to adverse weather conditions), vessels had to be 'fitted out', that is to say 'furnish[ed] with men, victuals and munitions for a period of twenty months'.[41] When ships returned, loaded with commodities, their cargoes were taken to the Company's warehouses at Deptford –, where the goods were controlled, inventoried, and sold. From 1648, after the Company had acquired East India House – formerly known as Craven House – on Leadenhall Street, ships unloaded directly in the City of London and commodities were taken to Leadenhall, or – space failing – to neighbouring warehouses.[42] As decades went by, East India House was remodelled and expanded on several occasions to create rooms matching its growing status. Ships continued to increase in size as well to accommodate larger cargoes. Some of the East Indiamen – developed in first half of the eighteenth century – reached the colossal weight of 1,400 tonnes.[43] The first steam ships started circulating in the mid-1820s and were making regular journeys to India the following decade.

Following the dramatic and unsuccessful Indian Mutiny, the East India Company was dissolved in 1858. The control and exploitation of Indian territories – and trade routes – were transferred to the British Crown. The opening of the Suez Canal in 1869 – and of the Pacific Railroad the same year – heralded the emergence of what historian James W. Frey calls a 'global moment', marked by the joint development of a planetary transport and communication network.[44] Faster steam ships, as well as the opening of a new maritime route, reduced the duration of the journey from England to

38 See Chatterton (2017 [1826]).

39 See Wild (1999), 67.

40 Ibid.

41 Chatterton (2017 [1826]).

42 See Wild 1999, 71. In the first fifty years of its existence, before the Company acquired East India House, the goods were stored across different warehouses across London.

43 Ibid., 74.

44 See Frey (2019), 9.

India by ten.[45] The shellac sold to record manufacturers until the Second
World War travelled on such steamers.

It is worth remarking that until the signature of the Declaration of In-
dependence in 1776, the US had no direct access to the global shellac trade
and was entirely dependent on Britain – with its worldwide monopoly on
products from the Far East – to secure the material. In the late eighteenth
century, the US established trade routes with India and started importing
the material in great quantities, paving the way for the mass manufacturing
of the first plastic commodities.[46]

Epistemic displacements and early scientific accounts

Like other colonial products including silk, shellac was first and foremost
a 'substance of speculation' – a term denoting both its financial appeal
and its broader 'potentiality'.[47] Accordingly, lac and shellac were coveted
for what they actually were as well as for what they might become. Their
appeal was poetic and scientific as well as commercial. On the one hand,
the evocative genesis and versatility of lac products constituted fascinating topics in and for themselves, arousing the interest of both specialised
readers and enlightened amateurs. On the other hand, these products
contained concrete promises of profitability and circulated through a
worldwide social and mercantile network, connecting the lives of highly
disparate individuals, including (to evoke but a few) cultivators, traders,
merchants, artisans, scientists, and genteel consumers. As well as entering
global commercial networks of exchange and consumption, the material
became part of a tightly-connected 'republic of letters', entering novel
epistemic networks – at a time when the realms of knowledge and commerce
symbiotically intermingled.[48]

As new stories of shellac were penned and circulated in the early modern
period, early indigenous treatises and folktales in Sanskrit or Sanskrit-based
languages were either ignored, neutralised or displaced (especially as only
very partial and occasional translations of these works were known to
Europeans at the time – most of these would not be translated into European
language until the second half of the nineteenth century and the peak of

45   Wild (1999), 82.
46   See Melillo (2020).
47   See Weszkalnys (2015), 617.
48   See Raj (2007), 16.

the orientalist craze).[49] This epistemic devaluation – which decolonial media and film historian Rakesh Sengupta describes as a 'subjugation of alternative knowledge systems and cosmology'[50] – was part of a broader colonialisation/modernisation process which valued Western epistemologies over non-Western ones. In particular, shellac's links with the feminine and cycles of life/death/renewal were gradually silenced; unsurprisingly, they are almost completely absent from eighteenth-century accounts. In other words, as it was imported, the material became neutral or neutralised. The material reappropriation of the substance was reinforced by modern scientific and/ or religious understandings, which largely uprooted existing indigenous interpretations. Accordingly, both the substance and its meaning underwent a first and substantial process of subjugation – and translation – as they travelled from the Indian subcontinent to Europe. Such a transposition, however, was not a one-way, linear operation. As Homi K. Bhabha eloquently suggests, '[t]ranslation is an iterative process of revision that moves back and forth in geographic circulation and discursive mobility, each time motivated by what is "untranslateable" – from one language to another, from one culture towards others – and therefore *must* be the cause for starting again from another place, another time, another history'.[51] While Bhabha largely insists on the emancipatory potentials of translation – and its crucial ability to *actualise* or *regenerate* meaning in and for the present – the 'displacement' he refers to (and the symbolic obscuring or erasing of an earlier stage) may also in some cases constitute an act of cultural violence, marking a place of erasure and forgetting.

From their inception, the East India Companies relied on local 'special-ized intermediaries' in order to locate and identify 'lucrative products, ranging from plants, herbs, and animals, to manufactured commodities'.[52] They counted amongst theirs ranks (and shareholders) a great number of scientists – specialising in botany, medicine or yet again navigational astronomy. In the eighteenth century especially, lac and shellac became vibrantly 'visualised' and depicted by a number of scientists, almost always

49 Some small proportion of this literature, however, was translated and available in European languages. For instance, the French surgeon Nicolas L'Empereur (1660–1742) – who found employment with the *Compagnie des Indes* in 1686 – engaged in an extensive and laborious work of translation of medicine treatises, palm-leaf manuscripts and books in Orya and other Sanskrit-based languages which he had translated into Hindustani, and which he subsequently translated himself from Hindustani into French (Raj 2007, 41–42).

50 Sengupta (2021), 3.

51 Bhabha (2018), 7; italics in the original.

52 Raj (2007), 19.

attached to the main East India Companies' expeditions. The botanical
and entomological literature of the eighteenth century counts many texts
dedicated to lac and shellac, and particularly to the description of the
lac insect – including the scientific reports of the French missionary Guy
Tachard (1709), the British surgeon James Kerr (1781), or yet again the British
naturalist William Roxburgh (1790, 1791). When Europeans discovered it, the
insect was frequently renamed after them, hence the names *Tachardia lacca*
(after Tachard) and *Kerria lacca* (after Kerr), a descriptor still widely in use
today. In the course of the eighteenth century, one also finds references to the
Kermes lacca or *Chermes lacca* – the term favoured by the British naturalist
William Roxburgh (who described it in the journal *Asiatick Researches* in
1791[53]). Other terms such as *Laccifer lacca* or the more generic *Coccus lacca*
were also used to reflect the insect's formal belonging to the larger family
of cochineal and other scale insects (such as the *Coccus cacti*, to be found
in Latin American, and yielding cochineal).

Tachard, Kerr and Roxburgh produced the first known scientific descriptions of lac and shellac production in European language (complementing
Linschoten's travel notes focusing on its applications), all of them drawing
from direct handling and in-situ observation of the insect. Many other
authors were content with recouping and replicating, with the occasional
flourish, existing descriptions. We find examples of such second-hand
descriptions in ambitious literary projects such as the French encyclopaedic *Dictionnaire des Origines* ('Dictionnary of Origins'), published in 1777
under Louis XVI. The dictionary sought to provide a complete survey of
the 'useful inventions' and 'important discoveries' of the time as well as
relating – amongst other things – the emergence of 'people, religions, sects,
heresies, laws, customs, fashions, money'. The entry dedicated to the lac
insect, though it derived its factual information from the scientific account
provided by Father Guy Tachard in 1709 (see section below), couched it in
a simplified and more candid, almost childish language – transforming a
scientific text into a mythological one. It generically portrayed the lac insects
as 'ants': 'The ants feed themselves from flowers, and because mountain
flowers are prettier and more abundant than seaside flowers, mountain
ants produce the most beautiful and reddest lac. Lac is to those ants what
honey is to bees. The ants only work eight months a year, and do nothing
for the rest of the time, because of the continuous and abundant rains'.[54] The
inaccurate yet evocative comparisons to ants and bees were not unusual, as

53 *Asiatick Researches,* published in Calcutta, was the journal of the Asiatic Society of Bengal.
54 Anon. (1777), 92; my translation.

they offered a familiar frame of reference for European minds, contributing to domesticating or Christianising unfamiliar insect species (both the ant and the bees, it must be noted, had strong biblical resonances).[55]

In the seventeenth and eighteenth centuries, the life cycle of the female insect (and its intrinsic drama) was recounted and further domesticated across a wide range of literary forms, including real or fictional travel diaries, scientific treatises and encyclopaedia entries. Such a phenomenon of 'graphomania' is not unsurprising in the cultural context of the time, notably characterised by the eclecticism of its print culture, its encyclopaedic drive, and the steady expansion of imperial powers (all three aspects being closely interrelated). Pre-nineteenth century accounts of shellac often drew from the same handful of primary sources; their authors thus replicated the same information (and inaccuracies) and were often ideologically biased. As we engage with such treatises, we become aware that, even in the Enlightenment age, scientific discovery was rarely decoupled from the mechanism and rhetoric of wonder – at a time when much of the scientific literature was written by religious men. Their texts betrayed a strong, content fascination with the indefatigable 'industrious virtue' of nature, its benevolent prodigality and its perceived 'miracles' as expressive of the divine spirit. Natural theological texts conventionally portrayed insects – such as bees, ants and beetles – 'as a celebration of God's powers in their alternative and surprising habits, singular instincts, industrious nature, and various forms of work. Insects and their ingenious and complex form of organization were sure proofs of the powers of the Creator'.[56] In other words, the organisation of the insect world mirrored, in miniature, the divine cosmos.

It is therefore no surprise that eighteenth-century Italian, Dutch and French Jesuit missionaries – such as Father Guy Tachard – should have been especially thrilled with the discovery of the tiny yet industrious lac bug, applying themselves to depict its life history in great detail. Tachard, who travelled to India no fewer than four times (dying there in 1712), became closely acquainted with its fauna and flora. He first described the insect in an important 1709 contribution, presenting the scientific community with the earliest account unequivocally demonstrating the function of the insect in lac and shellac production.[57] Before Tachard, the insect origin of the

55   See Melillo (2020).

56   Parikka (2010), 5.

57   See Parry (1935), 5. There are comparisons to be drawn between the cultural imagination and exploitation of shellac and the well-known eighteenth-century fascination with silk – a luxury thread originating from thousands of dead silkworm's cocoons (Anishanslin 2016, 1). Many of the treatises on sericulture were written by ecclesiastics – such as the Anglican Reverend Samuel

material – which was notoriously 'difficult to detect' given the small size of the parasite – had not yet been fully assimilated; it was often assumed that the shellac resin was derived from the tree.[58]

Reflectivity, imitation and intersensoriality

Scientific treatises on the material coexisted with a range of practical texts regarding its applications, highlighting its early adoption in the visual arts and its severance from direct practices of self-inscription. In European cultural practices, lac and shellac were never ingested or applied to the skin, but were involved in a wider visual culture of distanciation in which the body only played a minor and incidental part. Though it is impossible to precisely date the first time when it was used outside the South Asian area, occasional references to shellac appeared in Western texts prior to its more systematic and substantial exportation in the early seventeenth century. For example, lac was already a known material of painting in fifteenth-century Italy and, gradually, Holland and France. It is recurrently mentioned in the *Libro dell' Arte* (or *Craftsman's Handbook*) – a book of medieval 'recipes' and techniques compiled by Florentine painter and instructor Cennino d'Andrea Cennini around 1400.[59] Chemical analyses carried out in 1996 at the National Gallery in London testified to its actual presence in Italian medieval painting, confirming that painters such as Lippi (c. 1475), Ghirlandaio (c. 1480), Michelangelo (c. 1479), and Tintoretto (c. 1575) effectively had access to lac and knew how to extract the 'sanguine' red from the shellac resin.[60] However, no two shades of lac were ever quite the same, for its quality and tint depended (amongst other things) upon the particular host tree the female bug had fed from – a chemical variability it shared with the resin.[61]

Shellac was not a stable material. It could furthermore contribute to visually 'destabilising' other substances (including noble materials such as gold) – transforming and degrading them by contagion. As well as providing important insights into colouring materials, Cennini devoted a further section of his manual to the uses of shellac as a material of imitation. In a

Pullein and the Catholic naturalist Abbé Boissier de Sauvages. The silkworm's 'industrious virtue' inspired in them the same religious sentiment as the lac insect did. See Anishanslin (2016), 146.
58 Eastaugh, Walsh, Chaplin and Siddall (2004), 214.
59 Cennini (1960), 27.
60 Eastaugh, Walsh, Chaplin and Siddall (2004), 214.
61 Ibid.

section on the art of panel-painting and gilding, he explained how shellacked silver or tin could advantageously imitate gold.[62] Daniel V. Thompson, who established the definitive English translation of Cennini's handbook in 1933 and pursued his own material research into the realm of medieval painting techniques, further established that shellacked silver acquired gold's mirror-like powers of reflection.[63] Although not a noble material per se, shellac could thus persuasively imitate the soft, homogeneous brilliance of gold. Shellac and gold were also combined in another deceptive manner. Linschoten, reporting on his trip to India at the close of the sixteenth century, exposed the inexcusably 'cunning' practice of filling 'gold and silver works' with shellac in order to give the impression that they were entirely made of precious metals when, in reality, their cores were 'hollow'.[64] In its capacity of filler, shellac was classified by Linschoten as a non-substance, marking the empty place where the 'real', noble substance should have been. While he had praised the natural glass-like appearance of shellacked surfaces, the merchant rejected – on moral grounds – its uses as a hidden substrate, condemning it as a form of ontological deception and implicitly demonising the corruption of the indigenous people who practiced such ruse.[65] Incidentally, the practice of filling gold ornaments with shellac would become co-opted by European craftsmen in the following centuries.[66]

Although over one century separated them, both Ceninni and Linschoten (the practicing artisan and the erudite merchant) evoked the mimetic or transformational possibilities of shellac – glossing on its ability to become both a surface of projection and a substance of speculation (figuratively and economically). Shellac may indeed become gold. Because of its plasticity and lack of stability, it could potentially adopt one thousand shapes – anticipating the ubiquitous polymorphism of twentieth-century plastics so expressively depicted by Roland Barthes: plastic, rather than a single object, is 'the very idea of its infinite transformation'.[67] In the early days of synthetic plastics, this quintessential 'imitation material' was often 'aimed at reproducing

62   Cennini (1960), 88; see also Brill (1980), 108.

63   Thompson (1956 [1936]), 197.

64   See Parry (1935), 3. Linschoten made no distinction between lac and shellac – using the term 'lac' in a generic manner. I am re-establishing the distinction for the purposes of clarity.

65   Ibid.

66   Ibid., 173. This practice became especially widespread amongst jewellers in the nineteenth century. Examples of gold Victorian earrings filled with shellac can be found at the Victoria & Albert Museum in London, for instance. More generally, it may be noted that embedding precious stones in shellac was a widespread method of setting in India. See Osborne (1975), 459.

67   Barthes (1991 [1957]), 110.

cheaply the rarest substances: diamonds, silk, feathers, furs, silver, all the luxurious brilliance of the world'.[68] In the first half of the twentieth century, the mimetic properties of shellac would be further exploited by the growing movie industry, where it served (in conjunction with papier mâché) to rapidly fashion quantities of props and decors. Its high malleability meant that it could be used to imitate time-worn surfaces such as 'old plaster walls', 'oak doors' or 'iron work' and that it did so 'so perfectly as to defy detection'.[69]

From the outset, shellac constituted as much a material of preservation as it was of imitation or reproduction.[70] Its high inscriptibility – allowing it to reproduce extremely fine details – was recognised early. It had notably been involved in the composition of sealing waxes from the Middle Ages onwards – both in India and, later, throughout Europe and especially the Netherlands.[71] Emile Berliner would fully exploit the inscriptive proper-ties of shellac in order to store and retrieve sound waves in the mid-1890s. The entanglement of mimetic inscription and identity in relation to sonic etchings will be discussed further in Chapter 2, as we question more closely Berliner's claim that 'a modern disc record is a seal of the human voice'.[72]

Furthermore, the intersensoriality of shellac was important from an early stage. It is worth noting here that musical instrument makers in the seventeenth century already had a firm intuition of its acoustical properties. The practice of shellacking violins, violas and cellos started in Italy in the late 1600s and early 1700s – beginning with the instruments built by Antonio Stradivari and Giuseppe Guarneri.[73] The Cremona luthiers determined that a shellac layer, thinly applied to the 'carved spruce and maple sections of the instruments' bodies', would improve their resonance, significantly contributing to their distinctive bass tones.[74] It may be argued that the shel-lacking of instruments marked an intermediary and intersensory moment in the media history of shellac: a shellacked instrument acquired a *visual* as well as *acoustic* brilliance. Additionally, the reflectivity of shellac – and how it physically related to the 'brilliance' of sound – became an important

68   Ibid., 111.

69   Hicks (1961), 234.

70   Roy (2021b), 108.

71   See Sinha (1966), 4; see also Katz (1985), 8; Bartels and van der Hoeven (2005). The myriad shellac-based seals excavated in the Netherlands have allowed archaeologists to resurrect networks of correspondence and identities across eighteenth-century Europe, and to further theorise the different networks of 'contacts' which seals anticipated.

72   Hughbanks (1945), 19.

73   The practice of shellacking pianos occurred at a later stage. Shellac was also used to repair musical instruments. See Hicks (1961), 221.

74   Melillo (2020).

topic of discussion amongst the amateur recordists of the interwar period (see Chapter 3).

Toward the plastic century

Lac was only one amongst the many ancient dyes of animal or plant origins to be used across millenaries. It was often thought in conjunction with other insect-derived dyes such as cochineal and kermes (which, like lac, were all obtained from species of the *coccus* family). Amongst the red dyes, the best-known of them was certainly kermes, a dye originating from the Mediterranean area which was 'obtained from the dried bodies of female insects found in kermes oak'[75] and was used across Europe long before the introduction of cochineal and lac dye – both of which were discovered as Europe expanded and launched transoceanic expeditions. In the Middle Ages, cochineal – indigenous to Mexico and Central America (notably Brazil) – was first imported into Europe. It was chemically close to kermes, and was obtained from the *Coccus cacti*, a beetle feeding – as its name indicated – from some species of cactus plants. The *Coccus cacti*, yielding an intensely brilliant red pigment, remained a prized substance until the second half of the nineteenth century – only losing its appeal and centrality when the first synthetic dyes were synthesised. The lac bug produced a duller red than the brilliant scarlet obtained from cochineal, though its notable advantage was that it lasted much longer: as such, textile dyers tended to mix cochineal and lac dye in order to obtain a permanent, intense red.[76] The long and laborious process of dyeing textiles required highly specialised skills and knowledge of both natural fibres and dyestuff (for every colouring material, before it could be used, had to be separately processed and purified). Kermes and cochineal dyeing were amongst the first forms of dyeing to be successfully mastered and proto-standardised.[77]

The boom in synthetic dyes rendered both the natural dye trade and arduous processes of dyeing progressively obsolete in the second half of the nineteenth century. Following from William Henry Perkin's accidental synthesis of mauve in 1856 (as he was attempting to prepare quinine), a palette of synthetic colours obtained from an aniline derivative of coal tar

---

75 Brill (1980), 160.
76 See Hargitt (1923), 616; Anon. (1835), 116.
77 See Pastoureau (2011 [2008]), 32.

would become available, although a satisfactory red proved to be difficult to synthesise. A low-quality synthetic red dye was tentatively introduced in 1869, followed by another one in 1878 – known as 'ponceau', the latter was a strong scarlet which came to replace cochineal.[78] The synthetic dyes market soon took off, with research and production becoming almost exclusively concentrated in Germany where science and industry were more closely entwined than in England. Although the new dyes were initially of lesser quality than their natural counterparts, they were also far less expensive and much easier to procure.[79] Lac dye became permanently devalued as a visual medium. The development and proliferation of synthetic colours – and the organisation of the colour industry at the turn of the twentieth century – definitively halted the trade of kermes and *Coccus cacti*, devalorising those once sought-after commodities within less than five decades. This resulted in a sharp fall in import from India, with the Indian export data revealing that 'the export of lac-dye from India dwindled from 901.65 tonnes in 1868–69 to 51 kgs. in 1899–1900'.[80] At the same period, export of shellac 'shot up to 9918.7 tonnes in 1899–1900 from 2222.1 tonnes in 1868–1869'.[81] Were it not for its unique thermoplastic properties – accurately identified and exploited by industrialists in the second half the nineteenth century –, the *Tachardia lacca* too would have been completely phased out from Western industrial economies and discourses in the nineteenth century.[82]

The mid-nineteenth century was a strategic turning point for the global lac trade, marking a moment when the versatile resin became more valuable than the dye. As such, the modern plastic exploitation of shellac compensated for – or remediated – the obsolescent visual applications of lac. In time, the plasticity of the material would give way to – and authorise – its musicality or musical plasticity.[83] Shellac was first patented as a moulding material in 1854 by US photographer Samuel Peck in New Haven. In the early 1850s, Peck designed his first miniature cases – known as Union cases – to protect fragile ambrotypes and daguerreotypes from fading. Predating the mass circulation of gramophone discs by over forty years, Peck's glossy photographic cases, obtained from mixing shellac with wood flour, were

78   See Brill (1980), 165–166.
79   Ibid., 160.
80   Mukhopadhyay (2007), 29.
81   Ibid.
82   The term 'thermoplastic' refers to a material 'capable of being softened and made to flow by heat and pressure' (Gettens and Stout 1966 [1943], 332).
83   See Williams (2021).

'the first plastic products to be patented and mass produced in America'.[84] Although they were mass produced in Peck's Connecticut workshop, the miniature cases – and media containers in their own right – existed as singular 'objects of art', their daintily embossed covers channelling the creative sensibility of individual artist-engravers. Indeed, Union cases have often been hailed as paragons of nineteenth-century US design, with art historians praising '[t]heir rich heritage of designs [...] boast[ing] a greater variety than any other nineteenth-c[entury] craft'.[85] The plastic miniature cases were transitional objects – offering a bridge between artisanal culture and mass-production. As their popularity diminished in the late 1870s and processes of industrial casting and thermo-pressing developed, a new range of less sophisticated shellac-based commodities began flooding the market, seeping into the interiors and imaginaries of Western consumers. These early plastic commodities included toys and small dressing-tables articles such as combs, finely carved hand-held mirrors, bangles and trinkets, and were mostly aimed for women's adornment.

The mirror pictured below (Figure 2) depicts a bucolical scene set in the mythological past: a winged cherub whispers to a pensive young girl, languidly holding a freshly cut rose in her lap. Although the scene is silent, the cherub, carved in shellac, is speaking right into her ear – anticipating the voice of Berliner's 'recording angel' and the intimacy of mediated listening. Its hands are cupped so that no word is lost. A bird, its beak wide open, seems to be helplessly competing for her attention. Caught between the angel and the bird, the young girl doesn't seem to be listening. Her pose is one of listless abandon: she seems to be waiting for, or dreaming of, love.

Technically, the thermo-pressing of the first gramophone records relied on a process similar to the one used to mould shellac objects such as mirrors (though the grooves were less difficult to replicate than a fine drawing). Berliner and his contemporaries would have been familiar with such artefacts.[86] Fluid, malleable plastics offered a marked contrast with

---

84   Osborne (1975), 569. See also Katz (1995), 8. Peck's cases remained popular until the late 1870s, when they were replaced by plainer photograph albums.

85   Ibid.

86   Other usages could be mentioned here. The material was used extensively in the textile and particularly hat industry to help stiffen hats and other garments – and to make buttons. See Katz (1985), 8; Parry (1935), 171. In its mixed, liquid form (i.e., when the commercial shellac flakes were dissolved in alcohol), it was very well-known as a shiny protective polish for wooden furniture and constructions such as boats. In the 1904 children's book *The Merryweathers*, Laura E. Richards described the comings and goings of the shellac man, a teenage boy knocking from door to door to sell cans of pre-mixed liquid shellac in an American suburb. Shellac also served as an industrial coating and preservative for sweets, fruits and other foodstuff, giving them a

Figure 2: Back of an oval mirror of
black shellac, decorated with seated
woman and cherub in front of a castle,
nineteenth century. Science Museum
Group Collection © The Board of
Trustees of the Science Museum,
London

steel – the heavy, monumental material which, for Lewis Mumford, had
been the key substance of the previous 'palaeotechnic' era (c. 1750–1890).[87]
An iconic and unmovable steel landmark such as the Brooklyn Bridge
(inaugurated in 1883), whose engineering was so majestically evoked in
Alan Trachtenberg's material history, can be read as a culmination of the
palaeotechnic era, physically embodying as it did 'the forces, emotional as
well as mechanical, which were shaping a new civilization'.[88] By contrast,
the pliant and easily cast plastics announced a decidedly novel phase in
manufacturing – soon conjuring up a new material imagination. Even
though natural plastics were technically ancient materials of culture, their
potentials were more clearly perceived and profitably realised from the
1890s onwards.[89]

glossy, enticing appearance, while gardeners would apply it to trees to help repair damaged or
diseased barks.
87   In his 1935 classification, Mumford distinguished three main – and interconnected – phases
in the history of industrial civilisation, each of characterised by 'specific regions, resources,
raw materials, forms of energy, forms of production and workers' (Heath et al. [2000], 8).
88   Trachtenberg (1979 [1965]), 137.
89   Around 1900, many experiments took place to synthesise plastic materials, starting with
Bakelite (synthesised by Belgian-born chemist Leo Baekeland in 1907) which in the interwar would

By the beginning of the twentieth century, the trading groups of the seventeenth and eighteenth centuries – including the long-lasting English East India Company (which was dissolved in 1874) and the South Sea Company – had all disappeared, yet new transnational sites of knowledge transfer began consolidating. The 1920s and 1930s witnessed the emergence of a number of interconnected, often governmentally-funded epistemic and commercial sites such as the New York Shellac Research Bureau (based at the Polytechnic Institute of Brooklyn), the London Shellac Research Bureau (with offices in London and Calcutta), the United States Shellac Importers' Research Association, the Indian Lac Research Institute (founded in 1921 in Ranchi).[90] These constituted new networks and sites of knowledge transfer where skills were 'circulated, [...] constructed and reconfigured'.[91] The London Shellac Trade Association, established in 1931, gathered importers, dealers, and brokers who collectively regulated the shellac trade in London, from the moment when ships arrived and unloaded at Customs House.[92]

It must be noted that the various shellac associations mentioned above worked in close partnership with the consuming industries. The London Shellac Research Bureau and the British gramophone industry – the latter absorbing one third of the global annual output of shellac between the two World Wars – were frequent collaborators and research partners (with the Bureau's technical committee counting several members from the EMI firm, as indicated in the organigrams published in the Bureau's reports).[93] As well as undertaking research together, they were also engaged in wider dissemination activities, including exhibitions: they jointly curated an exhibit at the 1939 British Industries Fair, during which five thousand pamphlets on the history of lac and shellac were handed out to visitors.[94] In the first decades of the twentieth century, development in record-manufacturing processes were indissociably entwined with the research activities undertaken by the shellac industry 'in active co-operation with the consuming industries'.[95] It

become standardly associated with telephones and radios. See Katz (1985), 10. Plastic artefacts were multiple, anonymous rather than original, existing as dispersed and mobile emblems of modern manufacturing processes. The 1930s marked an intensification of the plastic age when oil displaced coal's dominance as the source of the chemicals, both contributing to reduce the costs of plastic artefacts and to stimulate the growth of new polymers.

90   The Indian Lac Research Institute was originally called The Indian Lac Association of Research and was renamed in 1925.

91   Raj (2007), 234.

92   See Parry (1935), 167.

93   See Mandal (n.d.), 34. DSIR 36/4271, TNA.

94   DSIR 36/4271, TNA.

95   Anon. (1936), 8.

is worth pointing out that the London Shellac Research Bureau was directly subsidised by the British Government. Firms such as EMI were, from their inception, implicit players in the broader geopolitical field: their continuity and stability – as well as their future – were entangled with the future of the shellac trade, and inexorably depended upon the maintenance of the colonial regime. The critical implications of this material and ideological entwinement would become especially manifest during the dramatic climax of the Second World War (see Chapter 4).

Bibliography

Anishanslin, Zara. 2016. *Portrait of a Woman in Silk: Hidden Histories of the British Atlantic World.* New Haven and London: Yale University Press.

Anon. 1777. *Dictionnaire des origines, ou époques des inventions utiles, des découvertes importantes, et de l'établissement des peuples, des religions, des sectes, des hérésies, des loix, des coutumes, des modes, des dignités, des monnoies, &tc.* K-M. Paris: Jean-François Bastien.

Anon. 1835. 'The Lac Insect (Chermes lacca)'. *The Saturday Magazine* 6 (175): 116.

Anon. 1936. *The Shellac Industry.* Namkum, Ranchi: The Indian Lac Research Institute.

Bartels, Michiel H. and Léon M. van der Hoeven. 2005. 'Business from the Cesspit: Investigations into the Socio-Economic Network of the Van Lidth de Jeude Family (1701–78) in Tiel, the Netherlands, on the Basis of Shellac Letter-Seals from a Cesspit'. *Post-Medieval Archaeology* 39 (1): 155–171.

Barthes, Roland. 1991 (1957). 'Plastic'. In *Plastics Ages: From Modernity to Postmodernity 1960–1991*, ed. Penny Sparke, 110–111. London: V&A Publications.

Baxter, Walter. 1954. *The Image and the Search.* Melbourne, Toronto, London: William Heinemann.

Berenbaum, May. 1995. *Bugs in the System: Insects and their Impact on Human Affairs.* Cambridge, Massachusetts: Perseus Books Blake.

Bhabha, Homi K. 2018. 'Introduction: On Disciplines and Destinations'. In *Territories and Trajectories: Cultures in Circulation*, ed. Diana Sorensen, 1–12. Durham and London: Duke University Press.

Brill, Thomas B. 1980. *Light: Its Interaction with Art and Antiquities.* New York and London: Plenum Press.

Bussabarger, Robert F. and Betty D. Robins. 1968. *The Everyday Art of India.* New York: Dover Publications.

Cennini, Cennino A. 1954 (1933). *The Craftsman's Handbook: The Italian "Il libro dell' arte".* Translated by Daniel V. Thompson. New York: Dover Publications.

Chatterton, Keble E. 2017 (1826). *The Old East Indiamen*. Project Gutenberg Ebook.

Eastaugh, Nicholas, Valentine Walsh, Tracey Chaplin and Ruth Siddall. 2004. *The Pigment Compendium: A Dictionary of Historical Pigments*. Amsterdam, Boston, Heidelberg, London, New York, Oxford, Paris, San Diego, San Francisco, Singapore, Sydney, Tokyo: Elsevier.

Engh, Barbara. 1999. 'After "His Master's Voice"'. *New Formations* 38: 54–63.

Fauche, Hippolyte. 1863. *Une tétrade ou drame, hymne, roman et poème traduits pour la première fois du sanscrit en français*. Paris: Benjamin Duprat.

Frey, James W. 2019. 'The Global Moment: The Emergence of Globality, 1866–1867, and the Origins of Nineteenth-Century Globalization'. *The Historian* 81 (1): 9–56.

Galvan, Jill. 2010. *The Sympathetic Medium: Feminine Channeling, the Occult, and Communication Technologies, 1859–1919*. Ithaca and London: Cornell University Press.

Hargitt, George T. 1923. 'Invertebrate Animals and Civilization'. *The Scientific Monthly* 16 (6): 608–622.

Heath et al. 2000. *300 Years of Industrial Design: Function, Form, Technique 1700–2000*. New York: Watson-Guptill Publications.

Hicks, Edward. 1961. *Shellac: Its Origin and Applications*. New York: Chemical Publishing Co., Inc.

Hughbanks, Leroy. 1945. *Talking Wax or the Story of the Phonograph*. New York: The Hobson Book Press.

Katz, Sylvia. 1985. *Classic Plastics: From Bakelite ... to High-Tech*. London: Thames and Hudson.

Knaggs, Nelson S. 1947. *Adventures in Man's First Plastic: The Romance of Natural Waxes*. New York: Reinhold Publishing Corporation.

Maillet, Arnaud. 2009. *The Claude Glass: Use and Meaning of the Black Mirror in Western Art*. Translated by Jeff Fort. New York: Zone Books.

Mandal, Jyoti Prakash. n.d. *A Study of the Problems and Prospects of Lac Industry in the Purulia District of West Bengal*. University of Burdwan, unpublished doctoral thesis.

Melchior-Bonnet, Sabine. 1994. *Histoire du miroir*. Paris: Hachette.

Melillo, Edward D. 2020. *The Butterfly Effect: Insects and the Making of the Modern World*. New York: Alfred A. Knopf. Ebook.

Miller Frank, Felicia. 1995. *The Mechanical Song: Women, Voice, and the Artificial in Nineteenth-Century French Narrative*. Stanford, California: Stanford University Press.

Mukhopadhyay, Asok. 2007. *A Study of the Shellac Industry with Special Reference to West Bengal*. University of Calcutta, unpublished doctoral thesis.

Osborne, Harold, ed. 1975. *The Oxford Companion to the Decorative Arts*. Oxford: Oxford University Press.

Parikka, Jussi. 2010. *Insect Media: An Archaeology of Animals and Technology.* Minneapolis and London: University of Minnesota Press.

Parikka, Jussi. 2015. *A Geology of Media.* Minneapolis and London: University of Minnesota Press.

Parry, Ernest J. 1935. *Shellac.* London: Sir Isaac Pitman & Sons, Ltd.

Pastoureau, Michel. 2011 (2008). *Noir: Histoire d'une couleur.* Paris: Editions du Seuil.

Raj, Kapil. 2007. *Relocating Modern Science: Circulation and the Construction of Knowledge in South Asia and Europe, 1650–1900.* New York: Palgrave Macmillan.

Roy, Elodie A. 2020. '"Total Trash": Recorded Music and the Logic of Waste'. *Popular Music* 39 (1): 88–107.

Roy, Elodie A. 2021b. 'The Sheen of Shellac – From Reflective Material to Self-Reflective Medium'. In *Materials, Practices, and Politics of Shine in Modern Art and Popular Culture,* eds. Antje Krause-Wahl, Petra Löffler and Änne Söll, 105–119. London, New York, Dublin: Bloomsbury.

Sengupta, Rakesh. 2021. 'Towards a Decolonial Media Archaeology: The Absent Archive of Screenwriting History and the Obsolete *Munshi*'. *Theory, Culture & Society* 38 (1): 3–26.

Serres, Michel. 1985. *Les cinq sens.* Paris: Grasset.

Sinha, Sukumar. 1966. *Census of India 1961: Handicrafts Survey Monograph on Lac Ornaments.* Calcutta: Government of India Publications.

Thompson, Daniel V. 1956 (1936). *The Materials and Techniques of Medieval Painting.* New York: Dover Publications.

Trachtenberg, Alan. 1979 (1965). *Brooklyn Bridge: Fact and Symbol. Second Edition.* Chicago and London: The University of Chicago Press.

Weszkalnys, Gisa. 2015. 'Geology, Potentiality, Speculation: On the Indeterminacy of First Oil'. *Cultural Anthropology* 30 (4): 611–639.

Wild, Antony. 1999. *The East India Company: Trade and Conquest from 1600.* London: HarperCollins.

Williams, Gavin. 2021. 'Shellac as Musical Plastic'. *Journal of the American Musicological Society* 74 (3): 463–500.

2.   Crackle: Assembling the record

Abstract:
Chapter 2 retraces how shellac progressively and predominantly became a medium of sound in the late nineteenth century, focusing on Emile Berliner's discovery of its sonic properties. The chapter especially focuses on the US where the material was progressively domesticated and 'Americanised', to the point of erasing its provenance and pre-mediatic histories. An important aspect of this chapter is that it makes visible the forms of labour entombed in the commodity of the record. It offers a parallel between shellac production in Indian workshops and the work carried out in western pressing plants, notably insisting on the crucial contribution of female labourers in the early phonographic industry.

Keywords: record industry, phonography, Berliner, shellac, labour, mimesis

Chapter 1 introduced the early uses of lac and shellac as visual media. It uncovered a first, pre-mediatic layer in the history of the material, charting the long period before it fully became a 'medium' (as well as an organised industry) in the modern sense of the term.[1] In this chapter, I continue to track shellac's transformative journey, examining another layer in the history of the material, and attending to its symbolic-material (re)emergence as a new (sonic) medium in the context of early phonography. This chapter focuses especially on the US where the shellac-based gramophone disc was developed by Emile Berliner in the second half of the 1890s. It asks how this arch-material of culture became distinctly modernised, domesticated and westernised (or even Americanised) through its sociocultural and industrial association with Berliner's gramophone disc. The latter, an easily reproducible media artefact, is often presented as single-handedly and magically 'pav[ing] the way for the development of a new consumer

1   See Parikka (2015), 37.

Roy, E.A., *Shellac in Visual and Sonic Culture: Unsettled Matter.* Amsterdam: Amsterdam University Press, 2023
DOI 10.5117/9789463729543_CH02

industry predicated on the mass production of music'.[2] But this technological
breakthrough simultaneously marked a moment of erasure. The early visual
stories of shellac were largely silenced and discarded as the 78rpm disc
began journeying into the lives and imaginaries of millions of listeners.
Other stories were encoded – or inscribed – into the material, with previous
understandings becoming increasingly faint and inaudible. The term 'shellac'
would soon become indissolubly entwined, in the cultural imagination,
with gramophone discs – and is still frequently used to this day (notably
by collectors) as a synonym for gramophone music.[3]

Despite its novel connotations as a material of sonic modernity, shellac's
legacy as a material of mnemonic inscription persisted. Berliner's personal
writings from around 1890 tantalisingly suggest that the invention of the
disc represented a poetic gesture of reinvention and recuperation – owing as
much to the scientist's concerns with improving sound recording technology
as it did to his fascination with antiquity. The advent of sound recording can
be understood psycho-archaeologically – as the crystallisation of a universal
archaic dream (that of immortalising the human voice and, by extension,
humanity itself). While this mythologising reading of technological innovation is attractive, it remains insufficient and flawed. The heavy human cost
involved in concretising such technological 'dreams' or, as Berliner called
them, 'wonders' must be acknowledged. The transnational phonographic
network, relying on a global infrastructure and organisation of resources,
cannot be detached from the larger extractive context which allowed for
phonographic supplies to be imported, economically appropriated, and (re)
imagined in the first place (as described in Chapter 1).

Processing shellac

Shellac remained a discrete yet key material of the transnational record
industry until the Second World War, with pressing-plants appointing agents
and testers to monitor its quality, as well as maintaining close relationships
with Indian suppliers and smallholders. It follows that the local, domestic
scale was nurtured and protected in order to prop up and sustain the global
scale.[4] At a very basic and minuscule level, early sound reproduction – and
the larger vitality of the global phonographic industry – can be connected

2   Jones (2001), 59.
3   See Petrusich (2014).
4   Smith in Herod (2011), 9.

to the reproductive cycle of the insect and to the perpetual, yet perpetually unpredictable, motion of natural cycles and seasons. The story of shellac combines the cyclical and the linear, the everyday and the historical – with localised routines feeding into larger histories of modern, multi-local capitalism. For Lefebvre,

> The cyclical originates in the cosmic, in nature: days, nights, seasons, the waves and tides of the sea, monthly cycles, etc. The linear would come rather from social practice, therefore from human activity: the monotony of actions and of movements, imposed structures. […] The antagonistic unity of relations between the cyclical and the linear sometimes gives rise to compromises, sometimes to disturbances.[5]

Shellac cultivation relied on a repetitive process and enlisted the energies of men, women, and children.[6] Four times a year, seasonal lac workers (*lahirís*) collected the whole branches or sticks on which the insect secretion was deposited. The main crop, *Baisakhi*, was harvested between April and July and accounted for over half of the total yearly production of shellac.[7] This 'stick lac', as it was called, was then crushed and graded: the largest particles ('seed-lac') would serve to produce the finest grade of shellac.[8] They were then washed (to remove the dye), placed in large baskets and separated by hand from their various impurities such as 'twigs, dead leaves, larvae'.[9] The sifting and sorting were principally allocated to 'nimble-fingered' women and children who, squatting around the baskets, worked quickly to remove the debris.[10] Once the first washing and sifting had taken place, the heated plastic mass was fed into long tubular canvas bags or stockings through which it was filtered, in order to remove the remaining, smaller debris.[11] This operation took place in the furnace room of the shellac workshop and involved two male workers: one experienced operator – known as the *karigar* – would

---

5   Lefebvre (2013 [1992]), 18.

6   My account of shellac manufacturing is based on reports from the 1930s and 1940s. It appears, however, that the manual process remained relatively unchanged through the centuries. One notable difference is that in previous centuries women were even more extensively involved in the process. In the second half of the eighteenth century, women operators were known to perform the role of the *karigar* and the *pherki*, and of the *bhilwaya*, as indicated in an 1876 report. See Parry (1935), 68–69.

7   See Adarkar (1945), 3.

8   See Gettens and Stout (1966 [1943]), 60.

9   Bell (1936), 28.

10   On the trope of nimble-fingered female workers, see Vágnerová (2017).

11   See Parry (1935), 65; Bell (1936), 28.

tightly twist and squeeze the bag while his assistant – the *pherki* (most generally a young boy) – checked that no impurity clogged its pores. Every now and then the bag would be slit open to remove the accumulated filth.[12] When it oozed out of the bag, the molten mass fell on a cold plate where it solidified in the shape of slabs. The operations would continue until the bag was empty.[13] The solidified shellac slabs were later warmed again by being deposited 'on to the outside surface of a porous jar containing warm water',[14] where they were pressed to form small squares. The squares were warmed in front of a fire and stretched 'with a rapid motion' by the male *bhilwaya* using his hands and naked feet to grasp their four corners – though rollers were also occasionally used to facilitate the stretching process.[15] The thin sheets (measuring approximately four feet square) were then left to cool down before being broken down into tiny shell-shaped flakes, ready to be commercialised and exported.[16] The thermoplastic flakes were re-melted and used in the manufacture of shellac-based commodities such as records. The trade requirements for high-grade shellac were 'freedom from dirt, insoluble matter [...] and lightness of colour'.[17] Although colour had no incidence on quality, record manufacturers privileged paler tones of shellac, setting up lightness as an industrial standard.[18] In order to 'improve' the colour of shellac and its marketability, toxic sulphide of arsenic (or orpiment) was often added to the seed-lac before it was melted and strained through the canvas bags.[19] The most widely traded grade was known as TN or 'truly native' (only handmade shellac would be awarded the TN denomination).[20]

12 These remainders would be recovered and sold as lower-grade shellac, known as kiri or passeva. See Parry (1935), 68.
13 See Bell (1936), 28.
14 Ibid.
15 See Bell (1936), 29; Gettens and Stout (1966 [1943]), 60.
16 See Gettens and Stout (1966 [1943]), 60.
17 Hicks (1961), 27.
18 Ibid.
19 Parry (1935), 60.
20 Shellac processing stirred the western imagination. For example, detailed descriptions of the pre-war and wartime shellac trade can be found in Walter Baxter's *The Image and the Search* (1954). The novel portrayed the professional and sensual life of a British woman owning a shellac factory (itself an unusual detail) in Bandhu, just outside Calcutta. It partly drew from Baxter's experience of Bihar. The feminisation operated by Baxter is a surprising feature of the novel, as well as its weaving together of shellac and romance. Williams ([2021], 485–487) gives us an example of a 1930 short story published in a London-based women's journal which similarly related shellac to romance. It is worth remarking that Baxter's novel was tried for obscenity – not because it graphically depicted the exploitative character of the shellac industry, but because it crudely portrayed the 'unprincipled' amorous life of an emancipated widow.

I would like to suggest that the establishment of a profitable record industry was partially authorised, in the first place, by the natural surplus and literal swarming of the invasive *Laccifer lacca* and its perceived inexhaustibility – as well as by the exploitation of a vast reservoir of readily available operators and experts. Shellac had been a significant sector of knowledge and activity for the Indian peasantry for centuries. In this regard, the survival of traditional epistemes was put to the service of a pioneering transnational sector of activity, with the early phonographic industry exemplifying 'the persistence of seemingly pre-modern means of production [...] in the smooth edifice of technological modernity'[21] and evidencing the different speeds and temporalities lying at the core of modernity, and by extension, of modern media assemblages.

One of the most complete and systematic portrayals of the lives of shellac workers appeared in a report of the Indian Labour Investigation Committee, put together towards the end of the Second World War and published in Delhi in 1945. The shellac industry operated as a loose, decentralised cottage industry: it counted over 350 factories in 1944, and about 5,000 *bhattas* (or furnaces), approximately employing 30,000 casual and seasonal workers (who were directly hired at the factory gates).[22] Neither wages nor working hours were regulated (payment was piece-rated), and shellac labour remained unprotected despite the existence of a Factories' Act (established in 1921). Burns frequently occurred from working near the open fire; workers' bodies were deformed from being cramped in unnatural positions for long hours, and some *karigars* developed blindness in late life from heat exposure.[23] The existence of the Workmen's Compensation Act was not well known amongst *lahirís*, and eventually there was 'no obligation on the part of an employer to pay compensation'.[24] A 1936 report on labour in India compared the conditions prevailing in feminised mica, carpet-making and shellac factories to those existing in the notoriously unsanitary tanneries and cigarette factories, highlighting the issues of child labour as well as the inadequate buildings, with their damp mud-floors where workers would 'sit or squat [...] throughout the working day'.[25] Despite the persistence and predominance of hand-processes, attempts at mechanising shellac

---

21   Melillo (2014), 235.
22   See Adarkar (1945), 9; 12.
23   Ibid., 6.
24   Rao (1936), 680.
25   Ibid., 678.

production intensified in the interwar period, most consistently under the impulse of the large British-run Angelo Brothers factory (located in Cossipore, in the industrial suburb of Calcutta).[26]

Domesticating shellac

In the late nineteenth, shellac was traded across the world and featured as a common material of US manufacturing – perhaps even as a quintessentially 'American material' and mythological product (in the Barthesian sense), forming the basis for the production of everyday moulded objects such as buttons, photographic cases or billiard balls. Demand for the substance would keep rising as western industrialists – and especially the nascent 'electrical and gramophone industries' – began mixing it with inert fillers to reinforce its surface hardness and discovered its advantageous thermoplastic applications.[27] Shellac was an inconspicuous household product, and its geographical provenance was progressively obscured. It may be that shellac was 'Americanised' in more than one manner, with several attempts being made at domesticating shellac cultivation – or at synthetising an efficient substitute for the resin.

In 1880, a certain Professor Stillmann gave a speech before the California Academy of Sciences, introducing an ambitious scheme to cultivate shellac in the more arid regions of the US (including the Mojave and Colorado deserts). The outlines of the scheme were duly reported in a *Scientific American* article titled 'Arizona Shellac'.[28] The motivations for bringing shellac cultivation in the US were of a predominantly pecuniary type, as it was believed that the US – which, in 1876 alone, had imported 700,000 pounds of shellac – could thus compete with British India, and procure work to thousands of unemployed 'boys'. Stillmann optimistically suggested that establishing a shellac industry in the US could be done with 'little or no capital'. It was soon found out, however, that the shellac industry could not be easily delocalised, as the reproduction cycle of the insect depended on the fulfilment of particular environmental conditions (including climatic

26   By the 1930s, Angelo Brothers company – first established in 1855 – had become 'the largest shellac producing unit in the world' (Anon. [1956], 45). The firm enjoyed close commercial relationships with British and US manufacturers, and would provide most of the British governmental shellac supplies during the Second World War.

27   See Bell (1936), xi.

28   Dated 10 April 1880, p. 225.

ones). The 'traditional technological knowledge' of workers could not be delocalised either.[29]

While shellac production could not be geographically displaced, it became obvious in the following decades that the material could be easily processed outside India. Important shellac-processing firms multiplied in Europe (particularly in Germany, which became the European centre of shellac production) and in the US in the second half of the nineteenth century. Among the big shellac-processing firms was William Zinsser & Company, founded in 1849 in New York by German emigrant William H. Zinsser (originally from Mainz). Zinsser's massive shellac factory, overlooking the Hudson River on 59th Street, appeared as 'a labyrinth of pipes and vats' strongly redolent with the sharp smell of 'alcohol dissolving the slabs of raw shellac'.[30] During the Great Depression, the factory still stood and flourished, and there existed several other shellac manufacturers across North America.[31] William H. Zinsser's great-grandson – the writer and biographer William K. Zinsser – affectionately evoked the peculiar joy he felt as he wandered with his sister Nancy through the entrancing shellac-processing facilities on 59th Street. Children of the wealthy Zinsser family embraced the lac insect as a quasi-member of the family circle: contrary to their contemporaries (for whom the origins of shellac were somewhat hazy), they 'knew from an early age the life cycle of the lac bug, which secretes a resinous cocoon onto the twigs of trees north of Calcutta' and 'never forgot that it was that humble insect that provided [their] privileged life and good education'.[32] Their loyal affections, however, flew most spontaneously towards the wonder of gramophone records. It may be no surprise that as an adult, having embraced a journalistic career, Zinsser went on to write extensively about recorded sound. Household names and brands such as Zinsser and Berliner progressively contributed to making shellac a familiar and domestic material. The absorption of shellac into US manufacturing

29   Smith (2015), 165. The attempt to delocalise shellac production recalls how, in the hope of reducing the cost of cochineal, colonial trading companies repeatedly and unsuccessfully attempted to transplant Brazilian cochineal and their host plant in various parts of the globe, including Australia and southern India. See Frey (2012).

30   Zinsser (2009), 156.

31   These included Bradshaw-Praeger, Haeuser, Gillespie, Rogers and Mantrose. These companies 'either imported shellac or prepared and packaged pre-mixed solutions of shellac and alcohol' (Mukhopadhyay [2007], 64).

32   Zinsser (2009), 156. The young William K. Zinsser – who was 'the namesake and only son' – refused to take over the family shellac business as was expected of him (Zinsser [2009], 16). Instead, he became a successful writer, teacher and journalist, one who 'processed' words rather than materials (Zinsser [2009], 155).

occurred through its absorption into everyday practices – and especially through the naturalisation of recorded sound – as well as through the cultural assimilation and valuation of figures such as Samuel Peck or Emile Berliner. Berliner's disc was patented in the US, a country where the figure of the inventor was invested with a special ideological potency. At the turn of the twentieth century, the celebration of inventors as 'national heroes' (or father figures) notably became a means of assimilating (or subjugating) heterogeneous immigrant populations into the common becoming of the nation – with inventors and their inventions serving the nation symbolically (by contributing to its self-mythologising narrative) as well as practically (by effectively increasing its material wealth, productivity, and cultural prestige on a global scale).[33]

Writing in 1926, Frederic William Wile – Berliner's first biographer – was keen to place Berliner within 'an aristocracy of inventive genius' who 'made rich contributions not only of American civilization, but of the human race'.[34] Wile's biographical narrative answered a patriotically-inflected agenda: its author sought to tell 'the story of the illimitable possibilities of America for the youth in whom the divine spark flickers, no matter how lowly or how alien his origin' through the paradigmatic 'story of an immigrant boy' – an *Hannoverkind* ('child' of Hannover). Wile made no difference between processes of individual and collective becoming (or between biography and history), and his biography can be read in light of Kasson's study of the messianic trope attached to new technologies and their individual inventors (such as Berliner and Edison but also, to a lesser extent, industrialists and entrepreneurs such as Zinsser). A techno-deterministic reading – such as the one offered by Kasson – suggests that Berliner's discoveries in the realms of telephony, microphony and phonography came to define and internally *organise* or 'wire' together, in the bio-neurological and technological sense of the term, modern America. In his recent analysis of late nineteenth century manufacturing in the US, Vaclav Smil similarly reactivates the trope of the philanthropic inventor, insisting on 'the role of individuals in the modernisation process', arguing that 'those were the decades of heroic invention and often admirably speedy innovation as engineers and entrepreneurs (often the same individuals) created a new world'.[35] My approach challenges such techno-deterministic claims. Although the following section gives a chronological and biographical account of Berliner's 'discovery'

33 See Kasson (1999 [1976]).
34 Wile (1926), i.
35 Smil (2013), 29.

of the shellac disc, it does so tentatively and does not posit that he was unilaterally responsible for the invention of the new medium.[36] The apparent linearity and sense of unity provided by the section below is destabilised and questioned by later sections of the chapter, which notably shed light on labour practices – and particularly racialised and feminised labour – in the early gramophone industry.

Experimenting with shellac: Berliner's wonders

Design processes (which are also living processes and processes of the living) are notoriously difficult to reconstruct and recapture, following as they do largely undocumented, untraceable mental operations and hazy, nonlinear trajectories (marked by heuristic moments of un-doing and erasing as much as by more measurable steps). Wolfgang Ernst, pondering upon what a 'media archaeology of one hundred years of radio (tubes)' may look like, reflected on the impossibility to produce 'an organized or even chronological history of the development of the tube, because the tube has no linear discursive history but instead, especially in the beginning, followed more of a zigzag course of experimental groping in the dark'.[37] Unsurprisingly, the development of the modern shellac disc – like that of the radio tube – proved long and arduous; it was guided by Berliner's indefatigable sense of experimentation and delight in materials as he tried making records out of substances as diverse as zinc, glass, copper, celluloid and rubber, before finally turning to shellac in the late 1890s.

A little-known unpublished essay entitled 'Wonders' and written in or about 1890 (as he was busy improving the composition of gramophone discs) gives us insights into Berliner's material imagination at the time. In this personal essay, Berliner affirmed that wonders were not to be found in 'the sphere of the supernatural, in the narrative of the Holy Scriptures, in the fables of antiquity, and in the séances of so-called spiritualism'.[38] Rather, he surveyed a series of 'everyday wonders', celebrating the vitality of substances as diverse as glass and musk – as well as evoking, in quick succession, the spellbinding laws of gravitation, magnetism, and electricity. In Berliner's lyrical prose, mechanical and chemical phenomena effectively begin to resemble magic, and something of the alchemist's exact enchantment

36   On the limitations of the biographical approach, see Kubler (1962).
37   Ernst (2013), 165.
38   Berliner quoted in Wile (1926), 33.

gets conjured up in his vivid descriptive vignettes. Matter appears to be potent, vibrant and sacred. His tacit framing of materiality as emergent and vivid, processual and animated, recalls Lucretius's theorisation of matter in *De rerum natura* – a treatise where the poet-philosopher described how 'the vitality of materials was inherent in the world's elemental structures and inevitably produced the collisions and accidents that affected history and social conditions'.[39] Berliner's predominant attitude remained one of reverent – yet scientifically informed – awe towards chemical and physical processes of transformation. He concluded his essay on 'wonders' with the suggestion that 'by teaching humility to its disciples, Science assumes the role of a most potent religion'.[40]

One may suggest that this humble, yet jubilant, mood of expectancy prevailed in Berliner's modest abode in New York, a city he moved to in 1875 after working for a year as a travelling salesman.[41] Berliner found employment washing bottles and chemical utensils in the laboratory of Dr Constantine Fahlberg,[42] an occupation which left him enough time to develop his independent theoretical study of materials and take evening classes at the Cooper Institute. There, he pursued his own interests in the realms of acoustics and electricity, finding enormous enjoyment in Johann Mueller's *Synopsis of Physics and Meteorology* (first published in 1854), which minutely covered Galvani's eighteenth-century discovery of fluid electricity and concisely summarised (in one chapter) everything that was known about electricity to this date. In 1876 – the year of the Centennial – Berliner began the naturalisation process to become an American citizen, moved back to Washington, and visited the Centennial Exhibition in Philadelphia – where he missed an opportunity to see an early exhibit dedicated to Alexander Graham Bell's membrane telephone. In Washington, his theoretical interest in electricity took a decisively practical turn as he began experimenting with batteries and wires purchased from the city's only electrical store. In 1910, in an autobiographical speech given at the Telephone Society of Washington, he fondly remembered the electrical supply store (owned by a Mr George C. Maynard) where he acquired his

39 Anderson, Dunlop and Smith (2015), 4.

40 Berliner quoted in Wile (1926), 336. The idea that science *was* a form of religion was strongly alive in Berliner. In the 1888 speech which accompanied the first public demonstration of the gramophone in Philadelphia, he had already evoked, in more contrasting terms, the dispute between science and religion. See Berliner (1888).

41 Most of the biographical details which follow are taken from Wile (1926).

42 Fahlberg, a Russian chemist who extensively worked with sugar, became famous for discovering saccharine in 1879.

early electrical equipment, soon transforming his rented quarters into a make-shift electrical laboratory.

At the same time, Berliner became fascinated with Bell's telephone, reading as much as he could about the machine (he had yet to see it) and noting the extreme weakness of its sound.[43] All throughout 1876–1877 he worked on the development of a telephone transmitter (or microphone) which would solve this particular problem, and drastically improve the viability of the machine. His microphone transmitter was to 'complete' Alexander Graham Bell's telephone[44] – a crucial improvement on the strength of which he was asked to join Bell's Telephone Company, relocating to Boston to work at the Bell telephone factory in September 1878. Once in Boston, Berliner carried on improving the telephone, largely contributing to the success of the Company. When the wax-cylinder graphophone was patented by Bell-Tainter in 1886 (itself an improvement on Edison's tinfoil phonograph), he naturally turned his attention to 'improving' the machine which (just like the phonograph) distorted sound. His gramophone was influenced by French printer and acoustician Edouard-Léon Scott de Martinville's early phonautograph (1857) which he had the opportunity to closely study at the United States National Museum in Washington on numerous occasions.[45]

The phonautograph mainly existed as a teaching aid to demonstrate the principles of acoustics. The device was used to write – and visually analyse – sound waves, yet it remained obstinately mute: it 'could only register sound, which was projected against a diaphragm and recorded on a moving cylinder around which paper covered with lampblack was wrapped. A lever or stylus was attached to the diaphragm, and this stylus traced the record on the smoked paper'.[46] Berliner assigned himself the task of making the machine talk. In order to do so, he built himself a small phonautograph, proceeding to make several 'pattern records' of his voice, wrapping strips of paper covered with soot around the cylinder, and fixing the voice-writings by pouring shellac varnish over them (this was the first time he used the material in connection to sound recording). Berliner next arranged to have some of these flattened paper strips photo-engraved onto zinc plates, enrolling the help of Washington photo-engraver Maurice Joyce (who, in the 1890s, had patented a new method for photographic

43 See Smart (1980), 425.
44 Although it featured a good receiver of sound, the early telephone had a poor transmitter.
45 The National Museum (1881–1911) was an early name for the Smithsonian Institution.
46 Wile (1926), 176.

matrices, and experimented with shellac).[47] He managed to replay the voice-writings manually, combining a steel pin, a sound box and the diaphragm part of a telephone receiver. Fortified by this early success, Berliner next sought to directly etch the voice into a flat surface (without having to resort to photo-engraving), and began experimenting with zinc plates covered with a soft layer of 'a fatty etching-ground'[48]– most often a mixture of beeswax and paraffin – to capture acoustic phenomena.[49] The engraved zinc plates would next be plunged into an acid bath to permanently fix the sound waves.[50] These prototypical zinc records could be played back and (imperfectly) copied through electroplating – yet the ragged, uneven grooves of duplicated zinc discs 'gave forth a sound that was somewhat raucous and harsh when compared with the sounds from a soft wax cylinder'.[51] Berliner therefore tried to find a more malleable material to reproduce records, notably experimenting with fragile glass discs and the more satisfactory celluloid – a horn-like, highly flammable synthetic plastic which had first been patented in the US in 1870.[52] Celluloid, which was too soft to withstand the weight of the playback stylus, was subsequently replaced by hard rubber (also known as ebonite, vulcanite or gutta percha).[53]

In 1894 the first batch of 1,000 gramophone records made with hard rubber were commercialised in Philadelphia, where Berliner had founded the Berliner Gramophone Company.[54] Complaints from buyers soon poured in: ebonite records bore 'flat spots' where the sound waves had not been properly etched; the stylus helplessly slid across these ungrooved areas,

47 See notably his 1893 patent entitled 'Matrix and Method of Using Matrices'.
48 Berliner (1895), 424.
49 Berliner first used expensive, pure New Jersey zinc. Dissatisfied with the crude results he obtained, he subsequently switched to cheap 'common zinc' (from Illinois, Belgium, or Silesia) which allowed for harder and thinner discs to be made. Common zinc had additional advantages over pure zinc; in particular, it 'took a very high polish, etched more readily, retained its flatness better – being somewhat elastic – and proved itself to be in no points inferior to the pure zinc, while it resisted better the wear of the reproducing stylus' (Berliner [1895], 424). Berliner also tried to replace zinc by other 'hard metals like copper, nickel, or brass' (Berliner [1888], 443).
50 These had been cut laterally rather than vertically, as was the case with the phonograph and the graphophone, thus drastically reducing distortion.
51 Smart (1980), 427.
52 See Berliner (1888), 437; Smart (1980), 427. It played a major part in the development of photographic and film stock – and in the manufacture of explosives. See Maxwell and Miller (2012), 72.
53 See Belchior Caxeiro (2021), 321.
54 See Smil (2013), 44.

causing a rupture in sound transmission.[55] Berliner quickly realised that hard rubber was not suited to reliably reproduce a matrix: it had to be given up. It is at this late stage in the history of the disc's development (the year was 1897) that Berliner turned, as a last resort, to shellac – a substance he had first used (without being aware of its sonic properties) to varnish his small 'pattern records' or voice portraits. He also briefly encountered shellac under the guise of an 'imitation rubber' composition in his work on the telephone, as he experienced with different materials for receivers. The material had favourably impressed him at the time, notably because of its handsome lustrous appearance and relative inexpensiveness (it was half the price of real rubber) – yet it revealed itself to be disappointingly brittle.[56] However, the shellac-based 'imitation rubber' proved to be ideal to press records for a number of reasons. Its first advantage was that it was 'easily moulded', and allowed for the 'delicate sound track on the matrix'[57] to be accurately reproduced. Furthermore, it was resilient enough to withstand the repeated pressure of the heavy gramophone needle travelling in the grooves, and sufficiently strong to be 'commercially handled'.[58] Its hardness also meant that 'the reproduced sound was louder and more crisp'.[59] Shellac is much tougher than wax and celluloid, with a remarkable abrasive resistance. Its durability made it 'the most suitable material for the production of records',[60] as it was considered that 'any first-rate record must withstand [the pressure of the gramophone needle] for at least one hundred playings',[61] after which the needle had to be replaced. Another practical advantage of shellac was that it was easily available in the US, especially on the East Coast (in addition to the Zinsser firm, other shellac providers in New York included Victory Shellac Works). Shellac subsequently became the binding agent of choice in the manufacture of 'millions of disk sound records',[62] soon to be adopted by all manufacturers: it is estimated that 90% of the records made before the Second World War were shellac-based.[63] In the US, up until the 1929 crash, the multinational import and export house Henry W. Peabody Co. – which

55   The flat spots were 'caused by gases developed by the rubber when heated' (Wile [1926], 200).
56   See Wile (1926), 160.
57   Bell (1936), 27.
58   Ibid.
59   Berliner (1913), 192.
60   Dearle (1944), 18.
61   Ibid., 19.
62   Wile (1926), 160.
63   Smart (1980), 427.

ran its own steamers to India – had a direct connection with the Angelo Brothers firm, supplying both the Columbia and Victor plants with the shellac they needed to manufacture records.[64]

The chemistry of 78rpm records

Berliner patented his gramophone in 1888, but it took him about a decade to formulate shellac-based discs, which were not mass-produced until the late 1890s, at which point shellac became a critical standard substance in disc-manufacturing.[65] For instance, a rare 1906 Pathé patent informs us that the firm's gramophone discs contained, at the time, 36% of shellac, 31% of kaolin, 22% of shale black, 7% of cotton and 4% of rosin.[66] In 1920, Columbia discs consisted of shellac, flux, paper flux, China clay, velvet black, smoke black, asphalt and methylated spirit (though the exact proportion of the different ingredients is uncertain).[67] Shellac mattered because it guaranteed the chemical integrity of the disc, holding the different components together. It was an essential *binder* but it was never the only component in the composition of discs, as a record made of pure shellac would have been far too brittle to be played – the material, used on its own, tended to crack.[68] Adhesives such as shellac fulfilled their potentialities in combination with other substances: it follows that gramophone discs were composite, eminently variable or unstable, objects, stemming from (varyingly successful) 'collaborations' between such heterogeneous materials as copal, silica, rosin and vinsol.[69]

Furthermore, although shellac was a standard material in record manu-facturing, the quality of crops relied on variable seasonal factors (including meteorological conditions). Record-pressing plants worked with and around such material instability – so that unpredictability itself paradoxically became a constant in the manufacture of records. The gramophone disc was never a fully 'standardised' or homogeneous product in the strictly material sense of the term, as testified by the circulation of countless 'disc recipes'. After Berliner evidenced the significance of shellac in disc manufacturing, numerous patents for improved 'recording plates' or 'recording tablets' were

64   See Zinsser (2014 [1956]), 73.
65   See Gilbert (2017), 3.
66   See Nguyen, Sené, Bouvet, le Bourg (2011), 34.
67   TNA LAB2/648
68   See Parry (1935), 71.
69   See Heath et al. (2000), 23; Anon. (1956), 94.

deposited in the US and Europe. Berliner's nephew – Joseph Sanders – was instrumental in refining record formulae; between 1908 and 1913, Sanders deposited five distinct patents regarding the production of records. His expertise would benefit the research departments of both the Victor factory in Camden, New Jersey and the Hannover pressing plant in Germany, the two principal laboratories where innovations in record making were developed.[70] In his 1905 patent, Sanders indicated for instance that shellac should be mixed with 'powdered barite or diatomaceous earth', in order to obtain a 'mass, which similar[ly] to the hard rubber softens in heat, but is very hard in cold condition'.[71] Sanders further remarked on the relatively high costs of procuring shellac, noting that disc-manufacturers were often inclined – or compelled – to use less shellac and adulterate their products, sacrificing the sturdiness and sound quality of the disc in order to maximise profits. Record manufacturers were even known to use scrap as filler in record compositions. Research carried out by the sound engineers of the National Library of Scotland confirmed that rubbish – including 'soft drink bottles litter, pieces of masonry and other unwanted pieces of material' – could be 'ground up together and mixed in' to make a new batch of compound.[72] As such, gramophone blanks – owing to their instable and even unpredictable composition – occupied a liminal and ambiguous position within the history of standardisation. On a superficial level, all gramophone discs bore a uniform appearance, overtly resembling one another (though a more attentive observer would easily detect differences in grain and shine, as well as in weight). On a deeper level, listening – as well as close chemical analyses – reveals a world of differences.

A number of archivists, conservators and sound engineers across North America and Europe have now begun tackling the monumental task of elucidating what gramophone discs may precisely contain, identifying and classifying the raw materials (or ingredients) used in record composition over decades. One of the aims of such chemical analyses is to be able to date the recordings with precision: as in archaeological excavations, the chemical makeup of materials may allow us to periodise artefacts. Indeed, a working hypothesis for conservators is that each type of disc composition may be mapped onto a particular period in record manufacturing, or onto a

---

70   These plants were respectively established in 1901 and 1898. See Belchior Caxeiro (2021), 329.

71   AT24624B.

72   As stated in an undated document internally issued by the National Library of Scotland. I am grateful to Nicola Reade for drawing my attention to it.

specific moment in the history of a record company. Records manufactured during the Second World War for instance, or in its immediate aftermath, can be easily identified in that they contain far less shellac due to the global shortage and governmental restrictions (see Chapter 4).

Furthermore, to determine the chemical composition of records also constitutes an important step towards knowing how to conserve them in the archive, and how to best shield sonic heritage from the ever-pressing threat of physical disintegration (and therefore inaudibility). Accordingly, research into the chemical composition of discs constitutes an important preliminary stage for digital sound preservation programmes across the world – and it may be argued that it is the prospect of digitisation alone which prompted many institutional sound archives to reengage with the hyper-materiality of their collections. In 2007, an interdisciplinary research project undertaken at the *Bibliothèque nationale de France* (French National Library) explored the physicochemical characteristics of the gramophone discs held in the Library's sound archive (which, with over 300,000 recordings, is one of the largest in the world). One of the aims of the research programme at the French National Library was to protect the collection from further physical deterioration, and thoroughly clean the discs before they could be digitised and preserved for posterity. The results of the research, involving a team of chemists and curators, were first published in 2011, shedding light on the bewildering range of materials used in record composition during the 1900–1928 period (with a focus on records manufactured by the Pathé, Zonophone, Gramophone and Berliner brands).

The proliferation of disc formulas is reminiscent of the multitude of recipes in medieval painting (see Chapter 1), pointing to the experimental and artisanal roots of early record production. In addition to shellac, the gramophone discs held at the French National Library contained nearly forty different types of substances.[73] While a small portion of these were man-made in a laboratory (record-plants appointed their own chemists),[74] most of these were sourced from the mineral and animal realm before being transformed, lending credence to the thesis of a 'natural history' of the gramophone disc. Some of these substances were used in record making very sporadically and tentatively before being abandoned (such as red ochre or

---

73   These included (in alphabetical order) asbestos, carbon black, casein, cement, clay, Colorado wax, cotton, flock silk, flour, kaolin, magnesia, mica, ochre, rosin or again Tripoli.

74   Including casein, an early type of synthetic plastic first produced in the late nineteenth century and derived from the milk protein, and largely used in the button industry. See Gilbert (2017), 4.

Colorado wax) whilst others (such as gum, kaolin and shellac) were present for years across almost all of the discs analysed by the team.[75]

Perfect records

Early and contemporary research into the chemistry of records has often been implicitly guided (or accompanied) by the fantasy of finding out what makes a 'brilliant' sound (the transfer of the term 'brilliant' from the visual to the acoustic realm is not accidental, and will be further discussed in Chapter 3) – where 'brilliant sound' becomes synonymous with 'perfect sound'. The idea of perfection and perfectibility runs through run through narratives of recorded sound to the present day – with each new cultural epoch fashioning its own myth of 'perfect fidelity'.[76] Such a fantasy of acoustic perfection existed long before audiophile equipment was developed in the 1950s and 1960s. Indeed, the early development of phonography was marked by a wish – on the industry's part – to 'improve' sound and 'perfect' machines. We may wonder what role materials and chemical assemblages play in aesthetic pleasure and audio fidelity. How is the experience of listening structured – enhanced, perturbed or diverted – by different types of material encounters? What can be said about the forms of affects which materials may enable, constrain or liberate? What is the chemical composition best suited to accurately reproduce musical compositions – and what happens when the two aspects are equated or put in relation? What sort of relationship is there between the shape of the record and matters of expressivity? And what is to be made of the disturbingly persistent 'material ideology' and format fetishism which have haunted audiophile practices and discourses ever since the advent of the tinfoil cylinder? It is generally understood that a smoother composition – high in shellac contents – leads to a higher fidelity in sound-reproduction because it minimises the sound of the medium. The proliferation of disc recipes – each of them seeking to make the mediating substrate inaudible – ultimately indicates the continually disappointed quest for a 'perfect' recipe, and by extension the failure of transparent communication.

75   The scientific investigation carried out at the French National Library is not a unique example: in recent years, similar research was carried out by sound archivists at the National Library of Scotland or by sound conservators such as Susana Belchior Caxeiro, examining audio carriers found in Portuguese phonographic collections. See Belchior Caxeiro (2021).
76   Devine (2013), 173.

It is worth turning here to Sybille Krämer's heuristic distinction between the postal and erotic concepts of communication. While the former is technical and emphasises the 'asymmetrical and unidirectional'[77] aspects of transmission, the latter is personal and is haunted by the (Romantic) dream of an unmediated, spontaneous dialogue of subjectivities (in romantic encounters, this is the typical fantasy of *total, mutual understanding* – understanding *without words*). In postal communication, the (material) media appears to be indispensably bound with the act of communication (implicitly, they become part of the message itself); in erotic communication, media constitute a potential obstacle or deforming mirror between beings and are therefore 'seen as detrimental to the immediacy of the dialogical'.[78] Here, communication defers communion. Indeed, Krämer, who wishes to '*rehabilitate the postal principle* and thus the transmission model of communication'[79] underlines that 'the more the materiality of the medium is shown to be technical, opaque, and compact, the more the notion of communication understood as dialogue [...] appears distorted'.[80] The two contrasting concepts are useful in the context of phonographic listening. On the one hand, listening – and especially the technologically mediated event of record-listening – constitutes a unidirectional and asymmetrical interaction where 'communication *lacks reciprocity* and is precisely not a dialogue'.[81] Yet by seeking to minimise and repress external parasites (including surface noise and natural distortion), an erotic and personal form of communication is sought. The dream of high fidelity – of pure sound – can be read in light of the postal/erotic divide. High fidelity represents an attempt to completely erase the sound of the medium (as with contemporary digital recording), that is to say to eliminate Krämer's fallible 'messenger' (or substrate) from the communication process, in the hope of achieving a seamless, intimate communion between self and sound, being and listening. In the realm of phonography, the playback device is supposed to efface itself in the moment of mediation and, indeed, to become completely inaudible.

It may be recalled that, in early phonographic demonstrations, Edison and his team of demonstrators insisted that their audiences should hear absolutely no difference between musicians giving a live performance, and the cylinder recording of their performance. Theirs was clearly a marketing

---

77   Krämer (2015), 22.
78   Ibid., 23.
79   Ibid., 24.
80   Ibid.
81   Ibid., 20.

exaggeration. In practice, sonic mediation was conspicuously noisy and disruptive. Early listeners of the phonograph were, in turn, entertained and outraged by the racket produced by talking machines. In 1888, after listening to tinfoil cylinders at a demonstration in London, a disgruntled reporter complained that 'their tone was metallic, nasal – sometimes a squeak, indeed – very often ludicrous or miserable'.[82] This was a common complaint. Hearing a cogent or 'perfect' acoustic message where there was largely noise was a leap of faith – requiring as it did considerable efforts or 'cultural work'.[83] When Berliner's shellac records were introduced in the 1890s, their thick surface noise similarly disrupted the experience of listening: the dream of immediacy was forever deferred, prompting Timothy Day to describe shellac recordings as 'extremely fuzzy snapshots, blurred round the edges, in parts indistinct and out of focus because of the limitations in playing the sounds back'.[84] In the early 1920s – before electrical recording was introduced – records were still perceived to be acoustically unsatisfactory. Compton Mackenzie – one of Britain's staunchest supporters of the gramophone[85] – saw recorded sound as an instrument for musical democratisation and education. Yet he complained that the dreamed-of perfection hadn't been reached, lamenting upon the 'devasting scratch' which drowned his otherwise wonderful records.[86]

Record listeners such as Mackenzie dreamt of perfect, immediate reproduction – they longed to silence the parasitic presence of the medium. Writing in the late 1990s and mourning the loss of (analogue) phonography as both 'a technology and social practice',[87] Eric W. Rothenbuhler and John Durham Peters emphasised the erotic principle of record-listening throughout decades, insisting that it was 'the possibility of perfection that motivate[d] the values of the audiophile'.[88] They argued that 'fidelity operate[d] as a transcendent value' which, as in a religious epiphany, brought listeners in contact with the 'other side', establishing a communion between 'the psyche of the listener and the stereo'.[89] The proximate, immediately enveloping sense of sound brings listening closer to the erotic concept (and, indeed, experience) of communication. Yet, because of its technological

---

82 *The Illustrated London News*, July 14, 1888, 30.
83 See Palmié (2019), 142.
84 Day (2000), 33.
85 Mackenzie was notably founded the *Gramophone* magazine in 1923.
86 Mackenzie (1925), 103.
87 Rothenbuhler and Peters (1997), 242.
88 Ibid., 253.
89 Ibid.

mediation, phonographic listening is bound up with the postal: the medium keeps reasserting itself (the hardware can fail). This unresolved tension between the technical and the personal may also account for the often described 'drama' of technologically mediated listening (phonography) or technologically mediated seeing (cinematography), hovering between presence and absence, emergence and erasure. Paradoxically, whilst praising the possibilities of perfection (or transparency) offered by phonographic listening, Rothenbuhler and Peters never ceased to celebrate the historicity of the phonographic medium: they underlined its indexical grain, opposing the physicality of analogue media to the perceived disincarnated existence of digital recordings.[90] The perfection of the phonographic experience, in other words, was derived from the very flaws and contingencies of the playback apparatus.

Likewise, it is the material heterogeneity rather than the homogeneity of the record compound which determines the smoothness of playback: in early phonography, perfect sound – or the perfect re-production of the voice – was therefore predicated upon the imperfect grain of the record. A perfect recording surface was produced from combining different elements together and securing the right 'mix'. The record was an exemplary site of material and technical hybridity. With its robust research and testing department, Columbia – the company which would patent the 33rpm long-playing record in 1948 – was one of the most proactive firms in the domain of experimentation (see Chapter 4). In 1922, the company began marketing their 'New Process' records, offering promises of 'a more silent surface': the disc ingredients were very finely grounded and sifted so as to erase, as much as possible, the grain of the medium.[91] Yet the process still depended on the collaboration of a great number of ingredients, which were refined to the point of dust. This 'miniaturisation' of the record particles in order to maximise sound definition is similar to what happens in digital photography where, the smaller the pixels, the clearer the image.

It can be noted that the ideal of perfection continues to haunt contemporary engagements with early phonography, surfacing for instance in the works of contemporary sound artists as they obsessively seek to create a materially 'perfect record' – a desire which can be compared to an erotic quest (in Krämer's sense). For instance, US sound artist and phono-archaeologist Robert Millis once wished to press a record out of pure shellac, undertaking

---

90   Ibid., 252.
91   Columbia had started experimenting with smoother surfaces in 1912. See Wilson and Webb (1929), 143.

research into Indian shellac factories in order to fulfil this ambition (see Chapter 5). Although the project was eventually discarded, its premise is interesting for it relied on the naïve (yet culturally persistent) fantasy that a pure shellac record would coincide with 'pure' sound. The quest for a perfect record remains utopian, demonstrating an anachronistic attempt at locating 'high fidelity' within a composite substance which, because of its natural origins, is bound to resist predictability, homogeneity and factory-like perfection.

Phonographic impressions: Masks and statues

Jacques Perriault once observed that the phonograph '[sat] at the confluence of the world of engineers and poets'.[92] In addition to its physical properties and commercial availability, it may be noted that the shellac material held an indefinable poetic resonance for Berliner. Remembering that shellac served as the basis for sealing waxes, he called the disc 'the seal of the human voice' while referring to the matrix as a 'recording tablet', by analogy with the portable writing tablets of antiquity.[93] The analogue media was thus seized through the inductive and connective mode of analogy, understood as 'the art of sympathetic thought' which 'forg[es] bonds between two or more incongruities and spanned incommensurables'.[94]

The phonograph's and gramophone's association with previous regimes of inscription – and 'metaphors of memory'[95]– was noted by both Edison and Berliner. If 'phonography maintains a certain form of intimacy with ancient forms of preservation apart from its status as an object of modernity',[96] it may be argued that this intimacy primarily resides in their common material substrate and in the specific operations of mark-making which this substrate authorises. For Berliner, to invent the grooved disc (as well as innovative ways of speaking through or about it) was also, in a poetic sense, to remember or recover its former shapes – to recreate the seal, albeit in the form of an acoustic seal. Analogical thinking emphasises spatial as well as temporal

92   Rothenbuhler and Peters (1997), 244.

93   See Berliner quoted in Wile (1926), 332; Berliner (1895), 424.

94   Stafford (1999), 10. Art historian Barbara Maria Stafford has retraced the history and political implications of analogical thinking from antiquity onwards, surveying how this once valued inductive mode was lost, falling into disgrace after Descartes and the separation of body and soul.

95   Draaisma (2000).

96   McCormack (2016), 179.

correlations: it weaves 'originality with continuity, what comes after with what went before, ensuing parts with evolving whole'.[97] At the turn of the twentieth century, the gramophone disc was discursively and physically resituated within a long genealogy of inscriptive devices and writing surfaces, through its tangible kinship with an archaic casting material.[98] In other words, the (apparently) new medium of recorded sound was mythologised (rather than historicised). Emancipated from the linear logic of progress, it thus acquired a hazily ancient, timeless resonance echoing Baudelaire's description of modernity as being half transient and half eternal.[99]

The mythologisation of the novel technical media pervading early accounts on phonographic devices can be read as a continuity of Baudelaire's influential coupling of the antique and the modern. In the 1860s, the poet and essayist was the first one to systematically explore the connection between modernity and antiquity – perceiving in the modern era seeds of the antique, and reciprocally. The 'antiquity' of the gramophone record figured prominently in Jean Cocteau's characterisation of sound reproduction, as sketched out in his 1930 diary. When the poet was first approached by Columbia to record his poems in the late 1920s, the prospect of the recording process led him to compare the gramophone disc to 'le masque antique' ('the antique mask'). Cocteau's mask can be read as both a theatrical prop and a mortuary mask (Giuliana Bruno, for instance, has evidenced the historical kinship between death and wax[100]). Coincidentally, a widespread, supplementary use of shellac in the first half of the twentieth century was to lacquer coffins. In this light, Francis Barraud's commercial paintings for His Master's Voice – showing the Nipper dog wistfully barking into a gramophone horn and sitting on the shiny, reflective tomb of his master – take on a new meaning. Phonography may be understood as 'a modernistic acoustic mask',[101] blurring the boundaries between the ancient and the modern, secular and religious thought. It is worth noting here that in the first decade of the twentieth century, acoustic transmission technologies (including telephony and phonography) were incorporated in a number of religious and spiritual experiments. Anthropologist Stephan Palmié has persuasively reconstructed

---

97 Stafford (1999), 9.

98 Many descriptions of early phonography belong to the analogical mode. It may be argued that the term 'phonography' itself established – at least nominally – a connection with former modes of writing and reading (even though the experience of recorded sound and its decoding widely differed from the deciphering of a written text).

99 See Baudelaire (1995 [1863]), 13.

100 See Bruno (2007 [2002]), 149.

101 Palmié (2019), 143.

how audio technologies became 'conduits for the numinous'[102] in a temple set up in Philadelphia by two Afro-Cuban brothers in 1908.[103]

Just like wax, shellac was an early medium of memory and mimetic impression: it could retain, for instance, the movement of the writing hand, or memorialise facial or other bodily features for posterity.[104] Wax had traditionally been the über material for memorialisation, preserving the memory of the living body – among its many memorialising uses, we can cite the wax-tablets so harshly decried by Plato in his *Phaedrus* (and further theorised by Freud), the full-sized wax sculptures common in Italian religious art of the sixteenth and seventeenth centuries, ex-voti, or yet again sealing waxes (often made from a mixture of wax and shellac). From early on, wax entertained a mimetic connection with the body and particularly to the externalised body of writing or mark-making. It was the medium which would aptly retain the 'forming' gestures of the artist, capturing the motions of the body and substituting itself to it at the same time.[105] Similarly, early direct recordings (such as the ones made on wax cylinders) contained a visual and sonic trace of the voice, petrifying a discrete part of the body and turning it into a discrete, autonomous, detached reality. Many early and contemporary accounts of phonography address this radical, literal disembodiment. Paradoxically the moment of recording (or dismembering) becomes the condition for further remembering and reassembling of the body.

As noted above, long before they became conventionally used in the making of musical records, wax and shellac traditionally constituted exemplary *recording* media (and media of inscription). Like Berliner, Cocteau drew attention to the disc's material affinities with masks as well as to (funereal) processes of casting.[106] He urged us to consider the ambiguous ontology of the recorded voice as simultaneously subjective and depersonalised, concealed and revealed, dead and alive, human and machinic, real and artificial, ephemeral and eternal – at once plastic and petrified (for the features of the mask, no matter how expressive, are rigidly set). An original aspect of

---

102 Ibid., 141.

103 It is worth remarking that the city of Philadelphia was an important place in the history of recorded sound, not the least because it was where Berliner had first publicly demonstrated the gramophone in 1888.

104 In a 2013 project entitled '*Un seuil*' ('A threshold') and realised in collaboration with Berlin-based sculptor Thomas Schelper, cultural historian of phonography Britta Lange rediscovered this use of shellac. For the use of shellac in the 'old art' of making mask, see Hicks (1961), 221.

105 Ibid.

106 See Cocteau (1999 [1930]), 213.

his thought was that it presented sound recording not as the production of a faithful acoustic double but as the fashioning of a third 'collaborative' voice between the human and the machine. The intermediary voice emerging from a mask (or *animating* a mask) effectively 'shifts the hierarchical relationship between subject and object'.[107] In the moment of mimesis, a form of contagion happens where 'erstwhile subjects take on the physical, material qualities of objects, while objects taken of the perceptive and knowledgeable qualities of subjects'.[108] The self and the machine meet (and merge) in the mimetic moment of (self-)recording, as recalled by early US home-recordist, clergyman and phonograph historian Leroy Hughbanks in his aptly-titled book *Talking Wax: The Story of the Phonograph* (1945). In such a context, artisanal home recordings (made on wax cylinders) may be further envisioned as auto-sculptures.

Anthropologist Michael Taussig defines the 'mimetic faculty' as 'the nature that culture uses to create second nature, the faculty to copy, imitate, make models, explore difference, yield into and become Other'.[109] The copy, he insists, absorbs the power of the original – it creates identity through a paradoxical process of estrangement and displacement. The 'second', a key notion in Taussig's *Mimesis and Alterity*, also plays a central role in Serres's reflection on statues. The philosopher proposes that a wax effigy comes into the world as a 'second body', that is to say as a substitutive body: it eventually comes to replace the first, original matrix (the copy stays whilst the biological body deteriorates and disappears).[110] The same can be said of a voice recording whose lifespan generally exceeds that of the individual who recorded it and which, metonymically, becomes re-membered (and acoustically re-experienced) *as* the person. Yet the 'original' and the 'second' do not exist in a linear or numerological succession: one does not precede the other, but they emerge simultaneously. The second retroactively creates the first.

In such a context, the origins of cultural production and transmission therefore lie with repetition. What Serres outlines here is a conception of culture as a material practice of re-production, where the past is continuously created through the process of re-tracing. For Serres, the quick of experience becomes translated – or 'hardened' – in the form of symbolic copies (statues) yet experience liquifies the statues. The instant creates the durable. The 'statue' (which closely outlines and embodies the real) must

---

107 Marks (2000), 141.
108 Ibid.
109 Taussig (1993), xiii.
110 See Serres (1989 [1987]), 220.

also be read in a less restrictive sense, as both an inscriptive machine and an inscription. It is an archival matrix (or, in Serres's words, a 'foundation') which gets re-mediated and reanimated in the moment of cultural encounter. On a phenomenological level, when we listen to sound recordings, the acoustic statue gets 'hallucinated' as a ghost.[111] After Derrida, audio-visual reproduction has frequently been theorised as the art of producing media phantoms. Yet the idea of the ghostly 'apparition' (defined by its brief and intermittent visits) is closely predicated upon the material survivance or persistence of phonographic statues. It follows that the former cannot be seized independently from the material conditions of its emergence. Similarly, the 'moment' of playback cannot be divorced from its long techno-cultural genesis.

Temporarily suspending the discussion on the ontology of sound reproduction, however, the final section of this chapter surveys how such 'statues' concretely came to life in the industrial environment of the early twentieth century.

Standardisation and industrial assembling

Though it was never materially stabilised or standardised, the 78rpm disc progressively emerged as a 'technical standard'[112] as well as a cultural standard in the first decades of the twentieth century. A clear emblem of modern manufacturing, and of mechanical modernity, it had become a 'worldwide standard' for sound-reproduction by the 1910s, progressively replacing wax cylinders as the listening medium of choice.[113] The two formats, however, would coexist until the 1920s. Disc reproduction notably relied on modern casting and pressing techniques which 'formed the core of industrial processes' in the first decade of the twentieth century, allowing for the rapid proliferation of 'exact replicas'.[114] As such, gramophone discs participated into the technology-driven culture of the copy identified by Walter Benjamin in his writings on mechanical reproduction (and particularly film and photography). The development of shellac-based records '*solved the problem of making unlimited copies of one original record*'[115] thus

---

111 See Ernst (2016), 117.
112 Tournès (2011), 16.
113 Ibid., 12.
114 See Heath et al. (2000), 23.
115 Reproduction had been a constant issue with earlier tinfoil and wax-cylinders.

'[laying] the foundations of a business of gigantic dimensions'.[116] Following a developmental phase of experimentation, the gramophone disc was
mass-produced in the late nineteenth century within a highly rationalised,
modern labour environment.

Although each manufacturer confidentially devised their own recipes,
the process of making discs was internationally standardised: the various
pressing plants across the globe relied on the same machinery – with presses
varying only in small details –, while operators were frequently given the
same set of instructions.[117] For instance, the British and British Indian plants
of the Gramophone Company were designed as mirrors of one another.
The first Gramophone Company factory was established in Beliaghata (a
suburb of Calcutta) in March 1907, the same year that the Hayes factory was
founded. It was moved to Dum Dum in late 1929. The Hayes and the Dum
Dum factories were similar both in terms of physical layout and internal
organisation. The Gramophone Company's Dum Dum plant was closely
modelled after the Hayes factory from the inside and shaped by colonial
contact. The building of the recording studio, for instance, was overseen
by a team of British technicians[118] while all the Indian employees received
direct training from British personnel of the Gramophone Company – so
as to absorb 'the easy flow of know-how from the parent organisation'.[119]

The early record manufacturers – such as the vast Camden factory in
New Jersey, the Dum Dum factory in British India or the Hayes factory in
Middlesex – might be naively glorified as 'truly nation-building', crystallising
'fascinating stories of quintessential human activities that created modern
societies'.[120] Yet the ideological insistence on 'modern wonders' – a category
which would also include 'rail, electricity, telegraph and telephone'[121] –
obscures the concrete conditions in which media technologies were actually
produced. Writing in the immediate aftermath of the Second World War, US
music critic Paul Affelder was aware that the effortless 'magic' of the 'shiny
black discs' blinded ordinary listeners to the 'great artistry and engineering
skills'[122] that were necessary to produce them. But stories praising the magic

116 Wile (1926), 201; emphasis in the original.
117 See Bell (1936), 33.
118 See Kumar De (1990), 24.
119 Ibid., 55. The relation of interdependence between the Hayes and the Dum Dum pressing
plants – with the former paternalistically dominating the latter – would carry on long after the
Partition.
120 Smil (2013), xi.
121 Maxwell and Miller (2012), 89.
122 Affelder (1947), 226.

of modern engineering are equally deceitful as stories celebrating the miracle of recorded sound, for they often focus on machinery, concealing the manual and mental gestures of the technicians operating it. Such a tendency to erase (gendered and often racialised) labour as well as resources from technological histories continues to prevail in contemporary accounts of new media.[123]

Archival research may assist us in our endeavour to recover buried labour practices in the record industry. In 1920, the British Ministry of Labour commissioned a report on the 'Gramophone and Musical Instruments Trade', offering us extensive and rare insights into the working conditions of those employed in gramophone factories after the First World War, shedding particular light on the work of women and young girls.[124] At the time, the entire trade counted about 150 firms in London and 65 in the provinces (most of these in Birmingham and Manchester). The five main record-pressing plants were located in the London area, the two main ones being the Gramophone Company (based in Hayes, to become EMI in 1931) and the Columbia Gramophone Company (London), both of which produced listening devices as well as records.[125] The Forbes report, as it was known, was based on interviews and direct observation: it began with the stern statement that '[t]he Gramophone and Musical Instrument Industry [was] badly organised'. One of its key findings was that, while both men and women were involved in the gramophone industry, pressing plants were 'run almost exclusively by women labour – unskilled girls are trained by the employers in the special processes required and consequently the employers are averse to any organisation dealing with wage agreements'. It follows that wages were extremely low (15/- to 25/ per week), workers were not unionised and no wage agreement existed because '[i]f it [was] discovered that a woman has complained to the Trade Union, she [was] immediately dismissed by her Employer, who at [that] time [had] no difficulty in obtaining female labour'.

The post-WW1 labour market was dominated by women who felt compelled to accept badly paid employment even though record making – an unskilled job – was 'most unsuitable and detrimental to their health'. Male

123 Though this has been progressively corrected in recent years. See notably Berthoud et al. (2012); Maxwell and Miller (2012); Nakamura (2014); Devine (2019a).

124 At the National Archives, the folder containing drafts of the Forbes report as well as summaries of their final findings is labelled TNA LAB/648. This section draws extensively from the data gathered by Forbes.

125 The three other companies involved in the manufacture of records were the Aeolian Company (London), J.E. Hough (London) and the Crystallite Manufacturing Company (Tonbridge) where female workers were 'employed on gramophone records, billiard balls, and buttons'. The Forbes report stated they '[had] long working hours and [were] poorly paid'. TNA LAB/648

workers were only involved in the initial phase of record making – known as the grinding process. The materials were weighed out and ground in heavy grinding rollers, and subsequently 'poured into an extremely complicated machine four stories high, known as a Banbury mixer. As they pass[ed] through the mixer, they [were] thoroughly mixed and heated, until they form[ed] a thick, black substance that greatly resemble[ed] molten lava"[126] and had a 'dough-like consistency'.[127] The pieced would then be 'rolled flat, cut into oblong pieces and baked"[128] by female operators and placed next to next to the press on a permanently hot table which ensured that they were soft enough to be placed into the press (which, in the US, was familiarly referred to as a 'waffle iron'). After folding it in half, the female employer would then proceed to place a shellac cake or biscuit (as they were known) – measuring approximately 10" x 5" – between heavy discs, before lifting and feeding it into the press – the discs, which were 'too heavy to be lifted by the arms alone', would be 'moved by being pressed against the body'. In the first phase of the process, steam passed through the press to keep it hot. The biscuit would 'melt and flow over the entire surface of the die, conforming to the mold created by the inverted grooves on the stampers'.[129] After fifteen seconds, cold water was passed through the record press which caused the record to harden and acquire its permanent shape (see Figure 3). It was placed on the side of the press to cool, after which its rough edges were trimmed by teenage apprentices ('the girls'). No waste was produced as the surplus material would be collected and placed back into the mix.

Discs were inspected before being collected and taken to the packing room, where they were packed by women and subsequently shipped via train to record storage facilities and dealers across the country.[130] The metaphorical, gendered language of baking so commonly used to describe the burdensome task of record making is worth noting: it deceptively lent record making an air of sweetness and domesticity, diminishing the efforts it took to obtain a 'perfect disc'. The workers were expected to turn out a 'perfect surface' as smooth as a mirror. Regular inspections took place both to monitor the visual finish of discs (which determined their sound quality) and ascertain when the stampers needed replacing (see Chapter 3). When the press was not hot enough, the record mixture (or 'dough', as it

126 Affelder (1947), 233.
127 Bell (1936), 31.
128 Affelder (1947), 233.
129 Ibid., 235.
130 Ibid. See also Devine (2019a), 73.

Figure 3: Industrial drawing of a gramophone record press designed by the Manchester firm Francis Shaw & Co. Author's collection.

was also familiarly known) would not flow properly, resulting in a defective pressing – a phenomenon known as 'cold pressing'.[131] Because their grooves were not properly formed, cold-pressed records yielded 'an irregular and rather disturbing kind of surface noise'.[132] When the press was too warm, the grooves were wavy, giving the surface of the disc an elegant sheen, very much like the shimmer of 'watered silk'.[133] Yet, no matter how attractive the defective records were, they would be rejected, their visual flaws instantly betraying their acoustic worthlessness.

It took approximately one minute to produce a record, and in 1936 a single press could handle about '500 records in an 8-hour working day'.[134] Fifteen years earlier, however, working days frequently exceeded eight hours and in 1920 the average working week in the gramophone industry was 48 hours (as evidenced in the Forbes report). Women would spend all day standing at steam-heated

131 Wilson and Webb (1929), 253–254.
132 Ibid.
133 Ibid.
134 Bell (1936), 34.

tables – with average temperatures ranging between 80 and 120° – in order to keep the shellac mixture malleable and soft. Their arms and elbows were 'constantly scorched and burned'. Pressing discs had a heavy physical cost, and most female workers were only able to perform the job for two to three months before 'leav[ing] broken down in health, and in many cases with their health impaired for life'. The reproduction of records (carried out in the matrix department) – relying on blackleading and electro-typing – was especially harmful as it involved working with cyanide of potassium, frequently leading to 'cyanide rash'.[135] The particularly harsh conditions prevailing in postwar British factories – especially for an unskilled female workforce – can be explained by the will to compete with US and German firms, and to generate quick profits (especially as firms such as the Gramophone Company had lost the important and profitable government munition contracts they had secured during the First World War; see Chapter 4).

In the US, record making was also a predominantly female sector of activity, with women respectively accounting for 80% and 75% of the workshop in RCA's Bloomington and Camden plants.[136] The Forbes report stated that record-pressing constituted a health hazard for women and flagged their (bodily, legal, and monetary) protection as an issue of 'national importance' – though no concrete steps were taken towards improving their working conditions in the immediate aftermath of the report. By 1930, however, the machinery was lighter and more automated, and it was possible for women to sit down while making a record (as shown by photographs taken at the HMV factory at the time). It must be noted that female labour was not confined to the sphere of pressing records but underpinned almost every aspect of the gramophone industry, including the assembling of gramophone cabinets and the manufacturing of motors and sound boxes. Producing horns was the only job almost exclusively undertaken by male workers.

Female operators have historically played a significant role in the production of early media commodities and technologies of telepresence (such as telephony and cinematography). In recent years, feminist film historiography has empirically readdressed 'women's labour in film and media production, whether in colour laboratories, cutting rooms, animation productions, service professions, or clerical and secretarial positions'.[137] A parallel work is

135 The dominant process to make a disc was known as 'the master and mother process' where a 'a recording blank [was] etched to create the matrix for a permanent mold or stamping for prerecorded discs' (Stauderman [2004], 31).
136 See Devine (2019a), 65.
137 Ozgen-Tuncer (2019), 273.

still to be done in the realm of recorded sound where the history of phono-graphic labourers remains largely unacknowledged (notably because it is still obscured by an insistence on individual inventors, recording artists and practices of consumption). It would be naïve, however, to assume that this history patiently awaits, ready to be magically or effortlessly retrieved from the archival matrix: the limits of archival data and historical artefacts, in the absence of oral histories, soon become palpable. And, as Cubitt remarked in his defence of anecdotal evidence, the dead will not be 'ventriloquised'.[138] The Forbes report quoted above represents a punctual, and in many ways exceptional, document which will need to be completed by, and contrasted with, a larger collection of sources (if and when these exist). The desire to address – or relocate – historically feminised labour practices is often linked, in varying degrees, to a contemporary feminist agenda. The latter should be openly acknowledged. What is sought here is not an authoritative 'rewriting' of (media) history but a reflection on (and against) erasing, an acknowledgement of the unwritten, the subjugated and the partially erased. While the lens of the political present may distort or 'tint' our views of the past, it is equally important to recognise that the (unsettled) 'past' may only be precariously fashioned from the (equally transient) ground of the political present. In this sense, there is no stable 'historical object', and historiographical practice always already exists performatively – as a continuous critical restaging of the past. It engages with the present as a moving horizon – which is continuously displaced or dislocated.

No full investigation of the labour conditions in the early phonographic industry presently exists, although important work has been made towards illuminating the supply chains and continued forms of financial and bodily exploitation underpinning the global electronics industry.[139] In particular, the gramophone industry may be (cautiously) re-interrogated in relation to Lisa Nakamura's analysis of early electronic manufacture, which she approaches as an 'integrated circuit' based on the cheap, flexible labour of Navajo women in(di)visibly connecting consumers and manufacturers of electronics.[140] Taking her cue from Nakamura, Vágnerová has portrayed the large contingent of female workers involved in producing electronics in the Third World, inviting us to consider what matters of labour may mean for media and music studies. She proposes that 'to account for the bodily harm of factory work is [...] not only ethical and critical, but also instructive

138 Cubitt (2020), 34.
139 See Nakamura (2014); Vágnerová (2017).
140 See Nakamura (2014), 919.

about music discourses and modes of listening'.[141] It may allow us to build another narrative of listening – or rather, to draw attention to the material and ideological conditions of emergence of modern listening practices (by paying attention not only to their hardware and technological mediations but also, before this, by asking how technologies get concretely sourced and assembled). Such a programme doesn't invalidate cultural (and/or musicological) analyses of individual media or musical artefacts, although it temporally takes us away from them, offering a form of 'musicology without music'.[142] The provocative phrase 'musicology without music', as coined by eco-musicologist Kyle Devine, operates as 'a call to expand and multiply' the understanding of music cultures by 'insisting that they are not only tied to but constituted by a variety of distributed and ostensibly non-musical people, things, and conditions'.[143] Such an approach insists on anonymous mediations (and pre-mediations), asking what makes cultural texts possible and transmissible at the most elemental, pre-mediatic level. Devine's claim that 'every system of inscription is tied to a system of extraction'[144] succinctly and efficiently highlights the interdependence of symbolic elaboration and resource extractivism in the context of recorded music.[145]

While this study similarly envisions phonography as a process of extraction (of resources, epistemes and energies) and impression/inscription (a literal 'marking' of the grooves, and of workers' bodies), it complementarily posits phonography as a particular system of expression (with records carrying and transmitting highly symbolic forms as well as materially indexing the history of their cultural formation). Keeping in mind these three entangled modes of phonographic existence – extraction, impression and expression – is useful to expand Devine's agenda, and to examine how these modes interrelate to jointly 'produce' the record. In addition to this, the production of phonographic commodities arose from a dynamic of inclusion and exclusion. Filtering occurred at every stage of the manufacturing process. At a basic level, manufacturers continually sought to

---

141 Vágnerová (2017), 252.

142 Devine (2019a), 21.

143 Ibid., 27.

144 Ibid., 24.

145 This resonates with Benjamin's famous words: all cultural treasures 'owe their existence not only to the efforts of the great geniuses who created them, but also to the anonymous toil of others who lived in the same period. There is no document of culture that is not at the same time a document of barbarism' (Benjamin [2003], 392). Inscription/extraction and culture/barbarism are dialectically and systemically entwined, existing in a relation of closeness and reciprocity.

determine what 'ingredients' should be included in disc recipes so that the commodity could function. The question of which repertoires were recorded – and, conversely, which sonic materials were marginalised and silenced – was further negotiated, on symbolic terms this time, by western record companies and sound archives in the first decade of the twentieth century (see Chapter 4).[146] In this sense, the activities of recordists such as Frederick William Gaisberg – who first visited British India in 1902 on behalf of the Gramophone Company[147] to record regional repertoires and develop the Indian market for recorded sound – could be compared to that of colonial traders, 'involved as they were in identifying, sourcing and buying the requisite raw materials'.[148] Processes of material and symbolic extraction, in the case of British India, closely coincided. More generally, selective practices of sound recording and sound archiving must be resituated within the broader colonial aspirations of the west, often inscribing themselves within the occidental model of appropriating and instrumentalising otherness. The ethnic groups and peoples whose songs were sonically captured in the early decades of the twentieth century were very frequently those who were colonised. Early phonographic archives and song-collection projects, relying on transnational recording expeditions, were often initiated with the hope of cataloguing and preserving all the musical traditions of the world.[149] We must continue to ask ourselves what kind of work and workers were made visible and what labour practices sand epistemes were, on the other hand, subjugated or glossed over. Which sections of the story were made inaudible and deactivated?

Recording artists were arguably the most directly visible workers or 'faces' of the phonographic network: but these visible faces had invisible, unpromoted counterparts. Ecomedia theorist Jacob Smith astutely proposes that 'work done in the forests of India' mirrored 'that done in American recording studios',[150] positing the operations of harvesting shellac and recording discs as interrelated or integrated nodes in what he terms the Green Disc network (and which echoes with Nakamura's 'integrated

146 The first sound archives were established in Europe at the turn of the twentieth century, respectively in Vienna (1899), Berlin (1900) and Paris (1911).

147 Or, as it was known at the time, the Gramophone and Typewriter Company Limited. The latter was renamed the 'Gramophone Company' in 1907.

148 Parthasarathi (2008), 29. On Fred Gaisberg's recording expeditions in British India in the first decade of the twentieth century and the industrial history of the gramophone in the Raj, see Parthasarathi (2008).

149 Fosler-Lussier (2020), 100–104.

150 Smith (2015), 24.

circuit'). While Smith correctly identifies shellac workshops and recording studios as two crucial laboratories of early phonography, his vision of the Green Disc network as a virtuous circle is a problematic one, for it doesn't acknowledge the deeper colonial context which brought India, Britain and, by extension, the US into such proximity. If one follows Smith's analogical reflection, the anonymous, repetitive manual assembling of records in the factory could be further contrasted with the culturally valued famous voices (and glamourous early recording artists) who symbolically produced 'standards' in the studio environment. It must be noted that the operations of recording and pressing records shared the same topography at Hayes: one section of the large Gramophone Company was occupied by the main, most sophisticated recording studios in Britain,[151] while another part of the site sheltered the record-pressing facilities. The geographical concentration of operations meant that new records could be manufactured and commercially circulated as quickly as possible after they were recorded. Famous recording artists and anonymous factory workers effectively shared the same physical environment, though their paths hardly ever crossed.[152] There also existed a deep yet unacknowledged solidarity between colonial shellac workers and metropolitan gramophone workers. Although they were arguably unaware of each other's existences, their respective and similarly precarious working and living conditions – notably in terms of high rates of female employment and lack of unionisation – bring them closer. These resonances should hardly surprise us, for shellac processing and record pressing happened on a geographically distinct yet economically and ideologically contiguous plane (that of global capitalism in the colonial era).

Phonographic palimpsests

As noted before, the history of phonography is a history of inscriptions and erasures: records themselves can be understood as palimpsests, teeming with

151 These had been established in 1912. The Company's first permanent recording studio was opened in 1898 in Maiden Lane, London. See Jones (1985), 82.
152 Clockwork motors and metal parts for gramophones were also produced at Hayes. See Martland (1997), 48. It must be noted that in its early years, the Gramophone Company did not manufacture listening devices or discs onsite: it only had a recording studio. It began pressing records in 1908, and to manufacture gramophones in 1912. See Jones (1985), 83–85. The Pathé Factory in Chatou (France) was organised according to the same model. Similarly, like its British counterpart, the Dum Dum factory manufactured gramophone machines as well as discs, and had its own recording studio. See Kumar De (1990), 21.

stories – not every layer is readily audible or retrievable. Some tracks remain temporarily or permanently muted; others become erased. The manufacture of records clearly relied on the exploitation of certain resources, territories, and bodies. At the same time, the disc produced its own territory – an actual as well as metaphorical topography. In the evocative technical language of the interwar period, gramophone discs were discussed as having 'lands' and 'grooves'.[153] A focus on the topography (or surfaces) of media objects attends to their scars, traces, and marks – interrogating what Cuban-born poet Octavio Armand described as the record's 'peculiar reality, the mechanical, uncertain nature of its incantation, its very awkward immanence'.[154] Armand poetically intuited that discrete scars acted as gateways into – and reminders of – deeper strata, materialising deeper accidents and specific histories of labour embedded within artefacts themselves (including visual histories of care and maintenance). It is obvious that the record, no matter how scratched, does not literally or magically show us the long history of its emergence. Narratives of cultural formation, however, may be materially deduced and unpacked by recovering the provenance of its components as well as their industrial and symbolic processing.

Accordingly, the visible topography of the gramophone record must be interrogated jointly with the deeper technical, cultural, social and political processes that relentlessly shape (and deform) its lands and grooves. The productive interplay (and potential disjunction) between media surfaces and infrastructures (or, perhaps more accurately, 'infra-structures') constitutes an important area of media enquiry: through 'unfolding the constitutive material layers' of media objects, 'political formations of labor, gender, colonialism, extractivism, ecology, and war' may become evidenced and better known.[155] In such a process of unfolding, however, monolithic abstractions – such as 'capitalism', 'modernity', 'music industry', 'power', or yet again 'empire'[156] –, get deconstructed. The material approach enables us to retrieve alternative and subjugated histories of early phonography. It allows us to address what novelist Georges Perec perceptively calls the 'infra-ordinary' and which, for the purposes of this study, could be termed the 'infra-phonographic'.

---

153 The term 'land' referred to 'the uncut space between grooves' while 'the grooves' were defined as 'the channel-like spiral track cut into a record by the stylus'. Anon. (1940), 120.

154 Armand (1994), 46–47.

155 See Devine (2019b), 19.

156 As observed by Radano and Olaniyan, 'The obsessive iteration of empire (as with a relative term, neoliberalism) leads to a loss of significance; its meaning comes to seem ambiguous, its power diminished' ([2016], 1).

Although media objects only indirectly speak for themselves, I believe they have 'a power to witness history that narratives lack'.[157] The role of media-material scholars may be to channel the witnessing power of objects in order to (re)convert materiality into narrativity – while recognising that not every story can be redeemed, retold or revealed. Objects – which often have a longer lifespan than human witnesses – may operate as interfaces or contact zones between different spatiotemporal sites. Memory objects such as sound recordings (even when they become partly inaudible) may remember us as much as they remember *for* us (in oblique ways). The attempts we make at translating – or transposing – things into words are not neutral and generic. Rather, they are necessarily original and situated: I see them as reparative and generative, offering us a tentative means of apprehending the contemporary moment through the fragile medium of the material past. This chapter has outlined ways of interpreting gramophone discs – which are specific media containers –, of letting them speak beyond their explicit or audible contents. Inspired by Perec's notion of the 'infra-ordinary', it has proceeded in an infra-phonographic manner, closely examining discrete surfaces to inventory voices and gestures that are usually overlooked, dismissed and excluded from histories of recorded sound.

Bibliography

Adarkar, Bhalchandra P. 1945. *Report on Labour Conditions in the Shellac Industry.* Delhi: Indian Labour Investigation Committee.

Affelder, Paul. 1947. *How to Build a Record Library: A Guide to Planned Collecting of Recorded Music.* New York: E. P. Dutton & Co. Inc.

Anderson, Christy, Anne Dunlop and Pamela H. Smith, eds. 2015. *The Matter of Art: Materials, Practices, Cultural Logics, c. 1250–1750.* Manchester: Manchester University Press.

Anon. 1940. *How to Make Good Recordings.* New York: Audio Devices, Inc.

Anon. 1956. *Shellac.* Angelo Brothers Limited: Calcutta.

Armand, Octavio. 1994. *Refractions.* Translated by Carol Maier. New York: Lumen Books.

Baudelaire, Charles. 1995 (1863). *The Painter of Modern Life and Other Essays.* Translated and edited by Jonathan Mayne. New York: Phaidon Press.

---

157 Marks (2000), 85. See also Taussig (1993), 232–233.

Belchior Caxeiro, Susana. 2021. *Immaterial in the Material: A Study on 78rpm Audio Carriers in Portuguese Collections.* Nova University Lisbon, unpublished doctoral thesis.

Bell, L. M. T. 1936. *The Making & Moulding of Plastics.* London: Hutchinson's Scientific & Technical Publications.

Benjamin, Walter. 2003. 'On the Concept of History'. In *Selected Writings, vol. 4, 1938–1940*, eds. Howard Eiland and Michael W. Jennings, 389–400. Cambridge, Massachusetts: Belknap Press / Harvard University Press.

Berliner, Emile. 1888. 'The Gramophone: Etching the Human Voice'. *Journal of the Franklin Institute* CXXV (6): 426–447.

Berliner, Emile. 1895. 'Technical Notes on the Gramophone'. *Journal of the Franklin Institute* 140 (6): 419–437.

Berliner, Emile. 1913. 'The Development of the Talking Machine'. *Journal of the Franklin Institute* 176 (2): 189–200.

Berthoud, Françoise et al. 2012. *Impacts écologiques des technologies de l'information et de la communication: Les faces cachées de l'immatérialité.* Paris, EDP Sciences.

Brill, Thomas B. 1980. *Light: Its Interaction with Art and Antiquities.* New York and London: Plenum Press.

Bruno, Giuliana. 2007 (2002). *Atlas of Emotion: Journeys in Art, Architecture, and Film.* New York: Verso.

Butler, Shane. 2011. *The Matter of the Page: Essays in Search of Ancient and Medieval Authors.* Madison: The University of Wisconsin Press.

Citton, Yves. *Médiarchie.* 2017. Paris: Editions du Seuil.

Cocteau, Jean. 1999 (1930). *Opium: Journal d'une désintoxication.* Paris: Stock.

Cubitt, Sean. 2020. *Anecdotal Evidence: Ecocritique from Hollywood to the Mass Image.* New York: Oxford University Press.

Day, Tim. 2000. *A Century of Recorded Music: Listening to Musical History.* New Haven and London: Yale University Press.

Dearle, Denis A. 1944. *Plastic Moulding.* London, New York, Melbourne: Hutchinson's Scientific and Technical Publications.

Devine, Kyle. 2019a. *Decomposed: The Political Ecology of Music.* Cambridge, Massachussetts and London, England: The MIT Press.

Devine, Kyle. 2019b. 'Musicology without Music'. In *On Popular Music and Its Unruly Entanglements*, eds. Nick Braae and Kai Arne Hansen, 15–37. Cham: Palgrave Macmillan.

Devine, Kyle. 2013. 'Imperfect Sound Forever: Loudness Wars, Listening Formations and the History of Sound Reproduction'. *Popular Music* 32 (2).

Draaisma, Douwe. 2000. *Metaphors of Memory: A History of Ideas about the Mind.* Cambridge: Cambridge University Press.

Ernst, Wolfgang. 2016. *Sonic Time Machines: Explicit Sound, Sirenic Voices, and Implicit Sonicity*. Amsterdam: Amsterdam University Press.

Fosler-Lussier, Danielle. 2020. *Music on the Move*. Minneapolis: University of Michigan Press.

Frey, James W. 2012. 'Prickly Pears and Pagodas: The East India Company's Failure to Establish a Cochineal Industry in Early Colonial India'. *The Historian* 74 (2): 241–266.

Gettens, Rutherford J. and George L. Stout. 1966 (1943). *Painting Materials: A Short Encyclopaedia*. New York: Dover Publications.

Gilbert, Marianne. 2017. 'Plastics Materials: Introduction and Historical Development'. In *Brydson's Plastics Materials. Eighth Edition*, ed. Marianne Gilbert, 1–18. Amsterdam, Boston, Heidelberg, London, New York, Oxford, Paris, San Diego, San Francisco, Singapore, Sydney, Tokyo: Elsevier.

Heath et al. 2000. *300 Years of Industrial Design: Function, Form, Technique 1700–2000*. New York: Watson-Guptill Publications.

Herod, Andrew. 2011. *Scale*. London and New York: Routledge.

Hicks, Edward. 1961. *Shellac: Its Origin and Applications*. New York: Chemical Publishing Co., Inc.

Hughbanks, Leroy. 1945. *Talking Wax or the Story of the Phonograph*. New York: The Hobson Book Press.

Jones, Geoffrey. 1985. 'The Gramophone Company: An Anglo-American Multinational, 1898–1931'. *The Business History Review* 59 (1): 76–100.

Kasson, John F. 1999 (1976). *Civilizing the Machine: Technology and Republican Values in America, 1776–1900*. New York: Hill and Wang.

Krämer, Sybille. 2015. *Medium, Messenger, Transmission: An Approach to Media Philosophy*. Amsterdam: Amsterdam University Press.

Kubler, George. 1962. *The Shape of Time: Remarks on the History of Things*. New Haven and London: Yale University Press.

Kumar De, Santosh. 1990. *Gramophone in India: A Brief History*. Calcutta: Uttisthata Press.

Lefebvre, Henri. 2013. *Rhythmanalysis: Space, Time and Everyday Life*. London, New York, Oxford, New Delhi, Sydney: Bloomsbury Academic.

Mackenzie, Compton. 1925. 'The Gramophone: Its past, Its Present, Its Future'. *Proceedings of the Musical Association* 51: 97–119.

Marchand, Eckart. 2015. 'Material Distinctions: Plaster, Terracotta, and Wax in the Renaissance Artist's Workshop'. In *The Matter of Art: Materials, Practices, Cultural Logics, c. 1250–1750*, eds. Christy Anderson, Anne Dunlop and Pamela H. Smith, 160–179. Manchester: Manchester University Press.

Marks, Laura U. 2000. *The Skin of the Film: Intercultural Cinema, Embodiment and the Senses*. Durham and London: Duke University Press.

Martland, Peter. 1997. *Since Records Began: EMI, The First 100 Years*. London: B. T. Batsford Ltd.

Maxwell, Richard and Toby Miller. 2012. *Greening the Media*. Oxford: Oxford University Press.

McCormack, Ryan. 2016. 'The Colossus of Memnon and Phonography'. *Sound Studies* 2 (2): 165–187.

Melillo, Edward D. 2014. 'Global Entomologies: Insects, Empires, and the "Synthetic Age"'. *Past and Present* 223: 233–270.

Mukhopadhyay, Asok. 2007. *A Study of the Shellac Industry with Special Reference to West Bengal*. University of Calcutta, unpublished doctoral thesis.

Müller, Martin. 2015. 'Assemblages and Actor-Networks: Rethinking Socio-Material Power, Politics and Space'. *Geography Compass* 9 (1): 27–41.

Nakamura, Lisa. 2014. 'Indigenous Circuits: Navajo Women and the Racialization of Early Electronic Manufacture'. *American Quarterly* 66 (4): 919–941.

Nguyen, Thi-Phuong, Xavier Séné, Emilie le Bourg, Stéphane Bouvet. 2011. 'Determining the Composition of 78–rpm Records: Challenge or Fantasy?'. *ARSC Journal* XLII: 27–42.

Ozgen-Tuncer, Asli. 2019. 'Historiographies of Women in Early Cinema'. *NECSUS* 8 (1): 273–281.

Palmié, Stephan. 2019. 'An Episode in the History of an Acoustic Mask'. *Archives de sciences sociales des religions* 187: 127–148.

Parikka, Jussi. 2015. *A Geology of Media*. Minneapolis and London: University of Minnesota Press.

Parry, Ernest J. 1935. *Shellac*. London: Sir Isaac Pitman & Sons, Ltd.

Parthasarathi, Vibodh. 2008. 'Not Just Mad Englishmen and a Dog: The Colonial Tuning of "Music on Record", 1900–1908'. *Working Paper* 2: 1–31.

Petrusich, Amanda. 2014. *Do Not Sell at Any Price: The Wild, Obsessive Hunt for the World's Rarest 78rpm Records*. New York, London, Toronto, Sydney, New Delhi: Scribner.

Radano, Ronald and Tejumola Olaniyan, eds. 2016. *Audible Empire: Music, Global Politics, Critiques*. Durham and London: Duke University Press.

Rao, Shiva B. 1936. 'Industrial Labor in India'. *Foreign Affairs*, 14 (4): 675–684.

Rothenbuhler, Eric W. and John Durham Peters. 1997. 'Defining Phonography: An Experiment in Theory'. *The Musical Quarterly* 81 (2): 242–264.

Serres, Michel. 1989 (1987). *Statues: Le second livre des fondations*. Paris: Flammarion.

Smart, James R. 1980. 'Emile Berliner and Nineteenth-Century Disc Recordings'. *Quarterly Journal of the Library of Congress* 37 (3–4): 422–440.

Smil, Vaclav. 2013. *Made in the USA: The Rise and Retreat of American Manufacturing*. Cambridge, Massachusetts and London, England: The MIT Press.

Smith, Jacob. 2015. *Eco-Sonic Media*. Oakland, California: University of California Press.

Stafford, Barbara Maria. 1999. *Visual Analogy: Consciousness as the Art of Connecting*. Cambridge, Massachusetts and London, England: The MIT Press.

Stauderman, Sarah. 2004. 'Pictorial Guide to Sound Recording Media'. In *Sound Savings: Preserving Audio Collections*, ed. Judith Matz, 29–41. Washington: Association of Research Libraries.

Taussig, Michael. 1993. *Mimesis and Alterity: A Particular History of the Senses*. New York, London: Routledge.

Tournès, Ludovic. 2011. *Musique!: du phonographe au MP3*. Paris: Autrement.

Vágnerová, Lucie. 2017. '"Nimble Fingers" in Electronic Music: Rethinking Sound through Neo-Colonial Labour'. *Organised Sound* 22 (2): 250–258.

Westermann, Andrea. 2013. 'The Material Politics of Vinyl: How the State, Industry and Citizens Created and Transformed West Germany's Consumer Democracy'. In *Accumulation: The Material Politics of Plastic*, eds. Jennifer Gabrys et al., 68–86. London and New York: Routledge.

Wile, Frederic W. 1926. *Emile Berliner: Maker of the Microphone*. Indianapolis: The Bobbs-Merrill Company Publishers.

Wilson, Percy and George W. Webb. 1929. *Modern Gramophones and Electrical Reproducers*. London, Toronto, Melbourne and Sydney: Cassell and Company.

Zinsser, William H. 1956 (2014). *A Family History and a Brief History of Wm. Zinsser & Co.* Self-published.

Zinsser, William. 2009. *Writing Places: The Life Journey of a Writer and Teacher*. New York: HarperCollins.

3. Mirrors: Phono-fetishism and intersensory visions

Abstract:

Chapter 3 surveys the intersensory position of recorded sound in the interwar period – also conceived of as a 'golden age' of shellac. It notably does so through recovering the largely forgotten – yet significant – trope of the 'mirror of the voice' and surveying how it was materially and discursively interpreted by groups as diverse as theorists, artists and home recordists. The first section discusses the visual phono-fetishism of the interwar period and critically reengages with Adorno's essays on phonography. As a counterpoint, the second part of the chapter attends to the defetishising discourse offered by interwar art and design practices (notably those carried out at the Bauhaus in Weimar).

Keywords: art, design, interwar, Weimar Republic, Bauhaus, phonography

At the turn of the twentieth century, shellac became primarily known and experienced as a material and medium of sound rather than of vision. The sonic and visual realms, however, were still entwined in more ways than one. While the first two chapters of this book scrutinise early trade networks, production and labour practices, this chapter engages with larger themes. It examines phonographic consumption and artistic expressivity in the interwar period, resituating phonography within larger historical debates on materiality and technological modernity. This deliberately broader emphasis also serves an important narrative purpose, mimicking the ways in which shellac became culturally absorbed in the sphere of everyday life between the wars (to the point of becoming invisible). The material momentarily disappears, only to reappear in unforeseen assemblages. I approach interwar phonograph culture as a fluid and multisensory object of study, wavering between the auditory and the visual realm. The interwar period, with its heightened intermediality and theoretical vitality, encourages us to explore

Roy, E.A., *Shellac in Visual and Sonic Culture: Unsettled Matter.* Amsterdam: Amsterdam University Press, 2023

DOI 10.5117/9789463729543_CH03

the changing and asymmetrical – yet not necessarily conflicting – relationships between different regimes of perception. It notably prompts us to
consider the vibrant persistence of vision within auditory knowledge, while
evidencing how the frontiers between the sonic and the scopic, rather than
being harshly defined, were always already porous and negotiable.

As well as introducing a thematic shift, this chapter also proposes a
geographical displacement. Moving away from Anglophone countries to
continental Europe, it principally engages with Germany and France where
phonographic consumption drastically increased following the introduction
of electrical recording in 1925. That year in Germany, the gramophone disc
became 'a true mass medium', with record sales running 'into six figures'.[1]
What had previously constituted 'a medium for connoisseurs and music
experts'[2] was now enjoyed by a continually growing circle of listeners
across a large variety of private and public sites. Phonographic listening
(and shellac consumption) intensified all across Europe – as well as in more
geographically and politically remote areas such as the Soviet Union. In
1927–1928 alone, Great Britain, Germany and France collectively consumed
18,000 tonnes of shellac to produce 260 million records.[3]

Changing consumption patterns were accompanied by debates regarding
the nature and implications of technologies in daily life. Theorists, artists,
designers, and scientists interrogated the ontological status of technological
devices while also experimenting with old and new materials to ask how
these might in turn alter or 'revolutionise' the everyday. For instance,
pioneering art, design, and architecture schools such as the Weimar-based
Bauhaus (founded in 1919) or the Petrograd State Institute for Artistic
Culture (founded in 1920) invited their staff and students to investigate the
turbulent chemical and physical lives of materials through tactile and visual
methodologies. The influential (yet, at the time still semi-confidential)
Bauhaus School of Design and Architecture,[4] set up by architect Walter
Gropius and staffed by many innovative material culture theorists and

1   Schrader and Schebera (1988), 117. The considerable technical improvement of sound-
recording technology – as well as the increasing affordability, compactness and portability of
listening devices – had a direct impact upon sales.
2   Ibid.
3   Berenbaum (1995), 123.
4   Bauhaus teaching combined the practical and the aesthetics, the mechanical and the sensual
in a multidisciplinary perspective (Turner and Badger [1988], 76). Gropius himself – in keeping
with nineteenth-century tradition, and especially William Morris's vision – initially believed
in the artist's power 'to breathe life into the dead product of the machine', proposing that 'the
sensibility of the artist must be combined with the knowledge of the technician to create new
forms in architecture and design' (cited in Naylor [1968], 34).

artists (including Moholy-Nagy), featured a plastic workshop and a laboratory for material thinking. In addition to working with traditional materials
such as glass, steel, and wood, Bauhaus tutors and students developed
research into new synthetic materials including, as designer Anni Albers
remembered it, 'light-reflecting and sound-absorbing materials'.[5]

Automation was another potent source of inspiration for artists, designers and theorists of the interwar period who addressed the particular
materialities – and peculiar animation – of machines. As Schrader and
Schebera commented in their exacting study of art and literature in the
Weimar Republic, 'Artists of all genres were fascinated and [...] influenced
by the new technological media of the twenties, and began to work for
them'.[6] Yet they did not blindly glorify or work for the technologies:
they also sought to work *with* them, patiently and often enthusiastically
uncovering their latent potentials. In many cases, they offered powerful
counterpoints to the pessimistic critique of technological consumption
charted by theorists such as Adorno and Heidegger. László Moholy-Nagy,
working at the Bauhaus, notably coined important theses on the radical
potentials of technical media of reproduction – including phonography – to
induce novel forms of experience and (by extension) of being-in-the-world.
What mattered there was to produce novel perceptions of the real rather
than merely copying it. It follows that a number of photographers and
filmmakers – rather than reinforcing the fetish of vision – used visual
media to subvert or decentre the eye: they notably did so by emphasising
the mobile, transitory and immaterial aspects (such as light, brilliance
and shadow) of images, and were especially interested in representing
movement. How can the writing of light (photo-graphy) capture – without
covering or erasing it – the writing of sound (phono-graphy)? Can these
two forms of inscription ever coincide?

Recorded sound, with its ambiguous, elusive materiality, was an especially
impossible – and therefore especially attractive – subject matter for visual
artists, driving them to test the limits and potentials of their expressive medium. Many artworks of the period are characterised by their intermediality,
conceived of as a process of inter-graphy: the production of an intermediate,
hybrid form of writing which heightens (rather than simply replicating it) the
heterogeneity of lived experience. This chapter brings together a selection of
visual works (originating more particularly from Germany and, to a lesser
extent, France) which actively and uniquely 'produced' recorded sound for

5   Cited in Naylor (1968), 122.
6   Schrader and Schebera (1988), 89.

the eyes to see – interweaving critical, poetic and playful perspectives. A central concept – and object – in this chapter is that of the mirror. Excavating a lost material and symbolic connection between records and looking glasses, I show how the mirror fantasy shaped the interwar imagination of phonography – surveying how it was extensively exploited by groups as diverse as commercial record companies, home recordists, critics, and artists. I further suggest that the analogy of the mirror can help us coin a different ontology of recorded sound. Indeed, a specular – rather than strictly spectral or hauntological – theory of phonography may offer us important ways of conceptualising the relationship between sound reproduction, identity and mimesis, as well as between listening and seeing. The mirror image contributes to partly defamiliarising the gramophone disc, subtly transforming (or destabilising) our vision of it.

Phonographic consumption and phono-fetishism

Berliner's shellac discs began to be mass-produced in 1898. Modern record-listening, however, did not consolidate until the interwar period and is notably to be linked to the introduction of electric recording in 1925, allowing for better, clearer and louder recordings to be made. There exist detailed sociological studies surveying the social and cultural history of interwar auditory cultures, describing its close entanglement with practices of mass-consumption, and mapping out the 'socialising function'[7] of phonographic objects.[8] Sociologist Sophie Maisonneuve has depicted the novel geographies, materialities and socialities of record-listening in the 1920s and 1930s – a period which could be described as a golden age for shellac and recorded sound. Her analyses focus more particularly on the genesis of gramophone clubs in the Great Britain and the rise of a male-dominated demographics of record collectors (also described as 'gramophiles' or 'discophiles'). In the 1920s, over 60% of British households were equipped with gramophones versus only one third in 1914.[9]

Maisonneuve has traced the collectors' hurried, hungry – yet disciplined – race for discographic knowledge and cultural capital, insisting on the primordial role of the specialised press in framing and cultivating the musical

---

7   Böhme (2014), 18.
8   See Maisonneuve (2002); Le Mahieu (1982).
9   Maisonneuve (2006), 28.

tastes of modern listeners.[10] In Germany, the 'Association of Record-Lovers', an offshoot from the popular Association of Book-Lovers, was founded in 1925 – coinciding with the introduction of electrical recording – and catered for a more diverse public than British listening clubs. Members of the Association could 'subscribe to discs from a number of companies at favourable prices'.[11] Collecting and consuming records became, reciprocally, a means to producing the self and, to some extent, a *standardised* (or perhaps automated) version of the self in the world. For Maisonneuve, musical consumption fulfilled a normalising or homogenising socio-sensory function, in so much as it helped fashion a form of trans-individual identity – based on the consumption of the same mediatic devices and contents – where the intimate and the collective psyches met. The eloquent oxymoron 'mass individualism' was coined by historian Moritz Föllmer to describe the increasing blurring of the private and the collective – and the convergence of social practices – which occurred during the Weimar Republic era.[12] Yet, despite the progressive consolidation and normalisation of phonographic listening practices in the 1920s and 1930s, the gramophone disc remained a symbolically instable, multivalent artefact, intersecting with a broad range of socio-cultural practices, beliefs and imaginaries.

An undeniable fascination with recorded sound and its material culture – which could be described as a form of phono-fetishism – consolidated in the years immediately following the First World War. The gramophone insistently pervaded the cultural imagery – and imagination – of the Weimar Republic and of the European interwar period at large. Cultural historian of phonography Stefan Gauss has coined the comprehensive term 'phono-object' to describe the myriad phonographic devices (including gramophones and radios) riddling interwar culture.[13] While he envisioned 'phono-object' as a neutral descriptor, this chapter introduces related terms such as 'phono-fetishism' to translate the uniquely modern and almost libidinal obsession with phonographic artefacts. According to Taussig, the 'representation of commodities in popular culture' was the place where 'the primitivism of modernity surfaced with unquenchable energy'.[14] It

---

10   Maisonneuve's study particularly focuses on the influential magazine *The Gramophone*, founded by Compton Mackenzie in 1923. By the close of the decade, the publication would count an average of 12,000 monthly readers, most of them male and belonging to the middle class. See Maisonneuve (2002), 49.

11   Schrader and Schebera (1988), 118.

12   Föllmer (2013).

13   Gauss (2017 [2014]), 73.

14   Taussig (1993), 231.

is worth noting that fetishism arose as a 'framework for interpreting European society'[15] in the nineteenth century, in light of the unprecedented proliferation and consumption of commodities made possible by industrial production. The 'phono-objects' of the interwar period, rather than being empty signifiers or indifferent decorative commodities, became both the medium and the source of a complex – and occasionally painful – cultural and ideological imagination. Cultural theorist Hartmut Böhme, in his substantial analysis of cultural fetishism, insists that fetishes are always 'material things', presenting themselves as 'matter that that incorporated something "other" than itself into itself: meaning, symbols, forces, energies, power, spirits, ghosts, gods, etc'.[16] Such an understanding of the fetish – rather than conveying conceptual rigidity – is particularly useful to attend to the plasticity of media objects and interpretative discourses, as well as to drawing attention to the persistent dialogue (rather than contradiction) between modernity and archaism.

As well as being visually disseminated in widely available newspapers and illustrated magazines,[17] phono-objects (encompassing gramophones as well as radios)[18] featured prominently in the realm of painting and more particularly in the disaffected still lives of the New Objectivists. Amongst the latter we can cite works such as Max Beckmann's *Stilleben mit Grammophon und Schwertlilien* (*Still Life with Gramophone and Irises*) (1924) and Rudolf Dischinger's *Grammophon* (1930). In these two paintings, the machine appears next to a plant – not so much to suggest a contrast between the machinic and the vegetal realms than to reveal the deep understanding silently passing between nature and technology – or the naturalisation of the latter. The German movement of 'Neue Sachlichkeit' (New Objectivity) – which developed in the 1925–1929 period – sought to answer 'a broadly felt need for clarity, sobriety, and stability'.[19] The New Objectivists helped make visible the striated, relational and fragmented nature of everyday environments – best understood as a succession of passing states – where objects and subjects were held together by fragile and mobile ties. Georg Scholz's *Weiblicher Akt auf dem Sofa* (1928) shows

15   Böhme (2014), 7.
16   Böhme (2014), 18.
17   These included the specialised journal *Phonographische Zeitschrift* ('Phonographic Journal') in Germany or *The Gramophone* in Britain.
18   For instance, the figure of the amateur radio hobbyist appeared in paintings such as Wilhelm Reise's *Verblühender Frühling: Selbstbildnis als Radiobastler* (1926) or Max Radler's *Der Radiohörer* (1930).
19   Hailey (1994), 16.

a naked woman propped up on a sofa covered with an oriental throw, pensively listening to a gramophone record, an opened box of colourfully wrapped chocolates by her side. The formal, academic genres of the still life and the nude painting were both reinvested by painters of the New Objectivity such as Dischinger and Scholz to accommodate hyper-modern artefacts, producing a slight effect of dissonance and anachronism – while retrospectively inscribing new machines within a conventional painterly continuum.[20]

Urban phonography

Practices of record-listening at home as well as in public spaces – such as parks, beaches and forests – were progressively intensified and normalised during the interwar era.[21] For Gauss, referring here to German auditory cultures, '[t]he filling of public spaces with sound from phono-objects happened as part of an increasingly dense, pluralized, and intense "acoustic environment", especially in large cities'.[22] Contemporary commentators noted, often with aggrieved resentment, the cluttered, cacophonic nature of urban soundscapes, often characterised by snatches of recorded sound escaping from open windows and noisily descending upon the street. Topics such as the blurring of the line between the private and the public sphere were heatedly debated in the regional and national press in specific relation to sonic privacy. In a bitter column entitled 'The End of the Public Sphere', published in the weekly magazine *Die Weltbühne* ('The World Stage') in January 1930, journalist Martha Maria Gehrke complained about the alleged end of quietness (idealising distant, pre-technological days). She saw herself as an involuntary 'pirate listener',[23] reduced to a state of phonographic martyrdom as she endured her neighbours' near-continuous listening sessions: 'If the radio is silent, then the gramophone resounds; there is no apartment house in which it would not be represented in numbers, no homeowner who lacks the altruistic need of allowing everyone around

20   Amongst the many paintings featuring sound-reproduction devices, we could also cite Hayden Henri's *Nature morte avec personnage* (1913), Max Beckmann's *The Night (1918–1919)*, Marek Włodarski's *Man with a Gramophone* (1926), or yet again Magritte's fantastical *The Menaced Assassin* (1927).

21   As depicted for instance in films such as *People on Sunday* (1929) or *L'Atalante* (1934). See Roy (2017) and Lewis (2015).

22   Gauss (2017 [2014]), 91.

23   Kaes, Jay, Dimendberg (1995), 613.

to take part in the perfection of his recordings'.[24] In 1922, Joseph Roth described a typical train ride past the houses bordering the S-Bahn line in Berlin. Peering into the interiors of his fellow city-dwellers, he noted how scenes of domestic listening visually and aurally bled into the public sphere, momentarily filling the space of the train carriage:

> A boy listens to a big phonograph on the table before him, its great funnel shimmering. I catch a brassy scrap of tune and take it with me on my journey. Torn away from the body of the melody, it plays on in my ear, a meaningless fragment of a fragment, absurdly, peremptorily identified in my memory with the sight of the boy listening.[25]

In his fictionalised memoir, British writer Christopher Isherwood similarly evoked everyday life – and memories of listening – in Berlin at the turn of the 1930s. *Goodbye to Berlin* (1939) was a prophetic title, announcing both the writer's imminent departure from the city and the forthcoming devastation of the great metropolis. The memoir gathers six distinct autobiographical fragments. It can be read as a literal chronicle or recording of time, inspired by the reportage style so favoured by German-speaking chroniclers of the period (such as Joseph Roth, Hans Fallada, and Siegfried Kracauer). Indeed, as Isherwood boldly proposes in the opening section ('A Berlin Diary'), there should be no difference between the eye of the writer and the passive lens of the camera. Revisiting (and updating) Stendhal's famous trope of the realist novel as 'a mirror that is taken along a path', Isherwood self-identifies as 'a camera with its shutter open, quite passive, recording, not thinking'.[26]

With Isherwood, Stendhal's nineteenth-century mirror is transformed into the recording machine of the 1930s. Meanwhile, the Stendhalian path has morphed into a busy city street. In keeping with its author's 'objectivist' writing principle, Isherwood's chronicles systematically encompass the lives of the living and the inanimate alike. In the age of objectivity, or 'New Objectivity' (as first defined in the aesthetic, pictorial realm by its 'politics of non-involvement'[27]), Isherwood tentatively presents himself as a blank carrier, indifferently exposed to the many scenes and sets of characters passing before him. He becomes the equivalent of a blank camera film or a yet unmarked gramophone disc – a depersonalised subject awaiting the

24 Ibid.
25 Roth (2003), 92.
26 Isherwood (1998 [1939]), 1.
27 Hake (1994), 129.

moment of contact in order to be 'sensitised'. Isherwood's conception of the subject as a passive blank – available to receive impression – contrasts with contemporary conceptions of the individual as always already wired, organised and heavily pre-determined (or hyper-coded) by technological mediations. For instance, Umbo's famous photomontage *The Roving Reporter* (1926) depicted the subject as the sum of its technological extensions (in a proto-McLuhan sense) – including a camera eye and a phonographic ear. Conceived the same year as Umbo's photomontage, Fritz Schüler's illustrated poster *Der Mensch als Industriepalast* ('Man as Industrial Palace') similarly offered an inner view of the human body as sheltering different communicating rooms of machines, each of them expressly controlling one bodily function. The living body was rationalised as 'an artifact or a technical system' which could be 'manipulated and repaired' – a view which, for Zielinski, prevailed amongst the mechanical, social and biological engineers of the 1920s – and was further encapsulated in Fritz Kahn's five-volume compendium *Das Leben des Menschen* ('The Life of Man') at the close of the decade.[28]

Rather than embracing the self as technologically organised and autonomised, Isherwood's novel expresses an ambivalence towards novel technologies, conveying the writer's uneasy relationship to the world he inhabits, and his estrangement from it. Isherwood writes, over and over, of loss and distance. This is less because he exists in exile, a young Englishman expatriated in a familiar yet culturally puzzling capital city, than because the real and the sensible world seems to be forever deferred, only accessible through the mediation of machines. For Isherwood, the city itself has become a threatening, gigantic machine. 'The Nowaks', another section of *Goodbye to Berlin*, reconstitutes the writer's days as a lodger in a shabby tenement building on the Wassertorstrasse in the eastern part of city. The whole architecture of the building is compared to a giant, enveloping listening device, with the shape of the court acting 'as gramophone horn', allowing the narrator to 'hear, with uncanny precision, every sound which came up from the courtyard below'.[29]

The rise of phonographic listening in cities – and the perception of urban architecture (from train carriages to buildings) as a gigantic sound-amplifying device – was further documented by the extensive photographic work of German press photographer Willy Römer (who ran the news picture agency Phototek), as well as by countless other professional and amateur

28   Zielinski (2018), 262. This mechanistic view of the living body was already prevalent in the previous decade (and particularly in the aftermath of the First Word War; see Chapter 4).
29   Isherwood (1998 [1939]), 148.

photographers. Römer, one of the most tireless diarists of everyday life under the Weimar Republic, published hundreds of photographs in the Berlin and the national press in the years preceding the Second World War. His subjects were varied, expressing the photographer's comprehensive, compassionate – yet rarely sentimentalising – understanding of his contemporaries' lives in the newly-established Republic. Römer's photographs give us a precise idea of the diverse social environments occupied by recorded sound. They portray modern beggars, pushing hand-cranked gramophones mounted on carts through the busy streets of 'Electropolis' – inscribing them within the long continuum of itinerant music-playing beggars.[30] They also chronicle the comfortable existences of children from well-to-do families, conscientiously practicing modern gymnastics whilst listening to (presumably) recorded instructions or fast-paced, rhythmical music emanating from the radio. Römer also documented, with a keen eye for the new and the uncommon, the practice of amateur recording, capturing young women visiting a fashionable 'Phonomaton' studio – or recording parlour – on the fashionable Kurfürsten-damm, recording vocal messages on thin plastic discs they would later mail to family members and friends. Mostly, he immortalised the interiors of the Berlin middle-class, depicting friends and families gathering around the gramophone, or couples dancing cheek to cheek to the sound of the radiogram.

Incidentally, the bourgeois apartments Römer kept photographing were not fully dissimilar from those occupied by a marginal 'New Soviet Middle Class' emerging in the mid-1930s in the Soviet Union – at a time when luxury novelties such as *patéfons* (Soviet hand-cranked gramophones), records, bicycles, cameras or yet again perfume, chocolates and champagne began being 'produced in their hundreds of thousands by Soviet factories'.[31] The example of the Soviet Union, as described below, offers us additional insights into the construction of phonographic commodities as fetishes – showing us parallel trends between the continental and Russian situation.

Sociologist Jukka Gronow has charted the notable – yet short-lived – rise of the Soviet consumer in the mid-1930s, attending in particular to the significant expansion of Soviet phonographic culture. He wrote about the erosion of the 'former ideals of revolutionary ascetism and social egalitari-anism'[32] as they gave way to more individualistic and hedonistic lifestyles inspired by Western trends. In such a context, *patéfons* were coveted as desirable symbols of status, progressively evolving 'from rare luxuries of

---

30   Their immediate predecessors had been organ grinders; see Silva (2019), 63.
31   Gronow (2003), n.p. The figures given below are taken from Gronow's study.
32   Ibid.

privileged consumption into items of everyday consumption'[33] as they reached non-urban areas. The purchase of *patéfons* significantly increased from 58,000 machines in 1932 to 675,000 in 1937, with a proportional growth in the production figures of records. Although the quality of Soviet gramophones left much to be desired – as revealed by the voluminous letters of complaint penned by disappointed customers –, its sheer existence meant that the Soviet government could 'boast of [...] closely following the latest achievements of technically more developed nations'.[34] The reproduction apparatus could therefore at the same time displace and re-produce – in a necessarily superficial, derivative manner – attractive foreign lifestyles. A patéfon placed 'on the table of a *kolkhoz* hut or in the summer garden of a *dacha*, with old and young gathered around it in a joyous mood'[35] served both to mimic a foreign reality and to domesticate (or Sovietise) the Western narrative of cultural progress. The gramophone in Russia had long been associated with the sybaritic unproductivity of the leisure class. Revolutionary poet Mayakovski, for instance, sarcastically itemised '[the] gramophone records, lace curtains, rubber plants, porcelain elephants and portraits of Marx in crimson frames' cramping the 'effeminate' interiors of the new Soviet middle class.[36] What the Soviet example magnifies are the perceived ideological affinities between the Western world (and especially the US) and the phonographic medium, so that recorded sound was still often approached as quintessentially American in the interwar years – a perception which also dominated in Germany (notably in relation to jazz recordings). In his private life, it must be noted that Willy Römer himself was drawn to the US – cherishing the country as a symbol of cultural openness and freedom – and it may be that his photographs helped artificially produce a faintly distorted, Americanised vision of German urban life – their documentary quality cannot be fully severed from the inner dreams and ideals of the photographer. However, and despite their Americanised filter, Römer's photographs partially translate a country-wide *Zeitgeist*, as well as giving us some insights specific to the Berlin culture of the time. They can be viewed alongside the contemporaneous photographs of his colleague Albert Renger-Patzsch, published in the 1928 photo-reportage *Die Welt Ist Schön* ('The World is Beautiful').[37]

33 Ibid.
34 Ibid.
35 Ibid.
36 Boym (1994), 8–9.
37 The photographer had originally (and revealingly) intended to publish under the title *Die Dinge* ('Things').

Technology and 'the realm of all'

Cultural fetishes, though they can be self-contained, never function in complete isolation but resonate with – and activate – one another. As such, the gramophone was not the only cultural fetish of the interwar period: it belonged to a broader network of strongly-connoted artefacts (which would also include radios and saxophone horns or yet again artificial silk – as immortalised in Irmgard Keun's 1932 novel *The Artificial Silk Girl*). The gramophone was a malleable, elastic symbol, associated with a multiplicity of meanings. Just as the ubiquitous saxophone evoked above, it symbolised an ambiguous desire for modernity and cultural otherness. It has been argued that Germany was intensively and intrepidly searching for new values after the trauma of First World War, placing great hopes in the future: as one contemporary critic put it, the country '[had] suddenly become endowed with an intense "modern consciousness" and look[ed] forward more eagerly than other nations because it [did] not care to look back'.[38] In such a perspective, technology would not only foster but also embody human progress.[39] Writing in 1930, the German-Israeli pianist and cultural consultant Leo Kestenberg affirmed, in his eloquent introduction to the anthology *Kunst and Technik* ('Art and Technique'), that '[t]echnology is the driving force of the era. It exerts a decisive influence on the profile of the times. Most of the phenomena which occur in the economic, social and political process are derived from consequences of technological development'.[40] He further proposed that technology constituted 'the realm of all', whilst art remained 'the realm of the few'.[41]

However, despite its democratic promises, 'technology' remained an ambiguous category, and was accordingly perceived in nuanced ways: an instrument of human progress and felicity could conversely become a repressive instrument of propaganda and control, symptomatic of an inhuman political regime (as was to be the case with the radio, which – since its early development before the First World War – was closely linked with the state and the military realm[42]). Consequently, while Römer and others wholeheartedly embraced the gramophone as a symbol of emancipation, the machine also came to represent a repulsive token of luxury and bourgeois

38   Cited in Turner and Badger (1988), 76.
39   Turner and Badger (1988), 76.
40   Quoted in Schrader and Schebera (1988), 89.
41   Ibid.
42   See Poschardt (1998), 43.

conspicuous consumption.[43] In the humiliating aftermath of the First World War, George Grosz repetitively used phonographic devices as a pictorial shortcut for US techno-cultural imperialism and oppression. For instance, his 1927 drawing *Deutschland, Deutschland über alles, über alles in der Welt (Diese Kriegsverletzten)* showed German crippled war veterans in tattered uniforms gathered around a phonograph blasting the German national hymn – nursing thoughts of revenge. His dystopian vision of technological progress echoed with that of American writer Henry Miller. Living in Paris at the time, Miller described the rising power of radiophony in his 1936 autobiographical novel *Black Spring* (written between 1934 and 1935). He denounced the tyrannic ubiquity of Bakelite radios in quasi-paranoiac terms, connecting the popular 'Song of Love' – a hit record of the day dominating the airwaves – to an implicit 'song of war':

> Out of the little black boxes an unending river of romance in which the crocodiles weep. [...] It is this *Song of Love* which now pours out of millions of little black boxes at the precise chronological moment, so that even our little brothers in the Philippines can hear it. It is this beautiful *Song of Love* which gives us the strength to build the tallest buildings, to launch the biggest battleships, to span the widest rivers. It is this song which gives us the courage to kill millions of men at once by just pressing a button. This song which gives us the energy to plunder the earth and lay everything bare.[44]

Miller presciently perceived in the radio an instrument of propaganda and large-scale affective manipulation. Listening as collective activity was envisioned as a dangerous endeavour, divesting people of their intellectual and emotional autonomy. By abolishing physical distances, and establishing new forms of intimacy, sound may weaken critical distance. The relationship between technologies of mass reproduction and power would come to be closely analysed by the theorists of the Frankfurt School. Of course, it may be argued that there was never a simple boundary. The emancipatory potentials of recorded sound were always inseparable from the threat of alienation. A technology which contributed to artistic liberation, self-awareness and social expression on the one hand could simultaneously serve and disseminate the discourse of barbarism on the other. The mobility of sound recordings and

43 It felt all the more scandalous considering the great financial distress of the Weimar Republic, its economy only slowly stabilising with the introduction of the Dawes Plan in 1924.
44 Miller (2009 [1936]), 154.

devices was not simply of a physical order: it could also be understood in terms of ideological mobility or reversibility. Sound could be turned inside out, over and over again: it was plastic in every sense of the word. From the disparaging drawings of George Grosz and the falsely candid documentary photographs of Willy Römer, through to the dreamlike, romantic intensity of Jean Vigo's *L'Atalante* (1934), painters, photographers and filmmakers contributed to visually producing, capturing and enhancing some of the values associated with the new technical media of the time.

The mirror of the voice: Self-fetishism and alienation

The above section has drawn attention to the visual ubiquity of phonographic devices, symbolically crystallising a number of cultural ideologies. In such a context, and as particularly expressed in interwar iconography, gramophones and discs manifested themselves as fetishes – shortcuts for the wider *zeitgeist* or mirror of an era. Complementary to this broad understanding of recorded sound as a 'fetish', a particular attention should be paid to the concrete association between gramophone discs and mirrors. In particular, the forgotten leitmotiv of the recording surface as an auditory mirror is worth excavating. The disc-mirror analogy constituted a recurring, prescient visual and narrative trope of the interwar period, permeating commercial visual culture as well as experimental art forms (and particularly photography and cinematography). In what follows, I retrace the links between sound-recording and mirroring – exploring not only their material kinships (see Chapter 1) but also the record as a place of projection, identity and potentially narcissistic self-possession. Accordingly, this section recovers the affinities between phonography and the specular, engaging with the gramophone disc as a material as well as a symbolic surface of projection and retro-projection. As well as punctually discussing the material properties of the shellac disc, I engage with Adorno's writings on the mechanical voice and the distorting mirror, discussing them in relationship to the later scholarship of literary and cultural theorists Charles Grivel, Felicia Miller Frank and Barbara Engh.[45]

Before attending to the trope of the mirror, it is worth pointing out that past and present cultural histories of phonography – and of mechanical music more generally – have conventionally highlighted the 'suspended animation' of recordings:[46] they insist on the spectral (or hauntological)

---

45   See Grivel (1994 [1992]); Miller Frank (1995); Engh (1999).
46   Rothenbuhler and Durham Peters (1997), 245.

dimension of technologically mediated sound, emphasising its mesmerising capacity to 'evoke *dead voices*'[47] and conjure up absent bodies. The desire to hear the past may even prompt us 'to hallucinate life when [we] listen to recorded voices', even though, in media-archaeological terms, 'we are not speaking with the dead but beholding dead media in operation'.[48] It may be noted that the idea of the ghost – and of immortality – was culturally embedded and encoded within the phonographic matrix from its early days. In their respective accounts of the new storage media, both Edison and Berliner – as well as the journalistic texts which they inspired – envisioned recorded sound as a means of sonically embalming the dead, or the future dead.[49] After the First World War, Edison further sought to produce a necrophonic device to communicate with the dead,[50] an imaginary machine anticipating Konstantin Raudive's capture of electronic voice phenomena (EVP) through radiophonic equipment in the 1970s.[51] Derrida links the realm of the spectral to that of technological reproducibility: anything which can be recorded and replayed – including sounds and images – anticipates an age of phantoms so that the cultural present (characterised by the proliferation of audiovisual traces) is also the realm of ghosts.[52] The modern era in this respect may be conceived as an exemplary site of haunting, with technologies of telepresence creating, storing and sustaining audiovisual phantoms. Reproducibility, as well as relating to the notion of immortality, anticipates what Pierre Nora termed an 'archival' form of memory which is predicated upon the stockpiling and retrieval of audiovisual traces.[53]

In addition to the well-known spectral reading of phonography, my suggestion is that the neglected *specular* mode offers us valuable complementary insights to understand the relationship between sound-reproduction and self-reflection as well as listening and seeing. Indeed, the dynamic terms of mirroring, recording, reflecting, and projecting may constitute valuable entry-points into the correlated poles of (re)production and repetition, identity and identification, self-knowledge and alienation. The triangular relationship between identity, mimesis and technologically mediated listening was most famously outlined by Adorno in his 1928 essay 'The Curves of

47   Miller Frank (1995), 146. On haunting and player pianos, see Ospina Romero (2019).
48   Ernst (2016), 117.
49   Sterne (2003), 289.
50   He explicitly and opportunistically had a market in mind, and believed that the families of the millions of fallen soldiers would be interested. See Baudouin (2015).
51   Citton (2017), 269.
52   Derrida (2010), 39.
53   Nora (1989), 13.

the Needle', the first of the three essays he dedicated to sound reproduction technology between 1928 and 1969. The author posited phonographic
listening as a potential source of alienation and self-objectification (or
self-fetishism), discussing what he pejoratively termed the (degrading)
'mirror function' of gramophone records (an analysis we will return to later
in the chapter), suggesting that:

> What the gramophone listener actually wants to hear is himself, and the
> artist merely offers him a substitute for the sounding image of his own
> person, which he would like to safeguard as a possession. The only reason
> that he accords the record such value is because he himself could also be
> just as well preserved. Most of the time records are virtual photographs
> of their owners, flattering photographs – ideologies.[54]

Though Adorno does not directly mention it, the idea of the gramophone disc
as a 'mirror' – and by extension of recorded sound as a visual reflection of
the self – constituted a common trope of the pre-war and interwar periods.
In a 1909 article which appeared in the Berlin-published *Phonographische
Zeitschrift* ('Phonographic Journal'), the writer Wilhelm Kronfuss compared
the gramophone to a variety of well-known domestic objects, including
'the optic lens, the telephone, or the flat mirror'.[55] 'Le miroir de la voix'
('the mirror of the voice') was, indeed, the suggestive motto coined by the
French record and film company Pathé Frères (founded in 1896) to promote
the brightness and realism of their recordings.[56] In the 1920s, these were
sold in ordinary blue or brown sleeves – and were especially sought after
by French and Russian listeners. The left-hand side corners of the sleeves
were stamped with the stylised monochrome drawing of a short-haired,
emancipated flapper. She holds a record to her face as one would casually
hold a looking glass, closely inspecting her reflection – so that the disc
becomes a portrait in sound or a 'sounding image'.[57] The Pathé logo functions
as a visual pun, delivering a seemingly simple and efficient message: in
highlighting the sound recording's perfect verisimilitude and closeness to
real life, it visually anticipates the concept of high fidelity (a term which

54   Adorno (1990), 54.
55   Gauss (2017 [2014]), 74.
56   Pathé records were unusual, vertical-cut records (as opposed to the widespread lateral-cut
discs), and relied on a hybrid technique of recording, 'a sort of compromise between the Edison
and Berliner methods' (Gronow and Saunio [1999 (1998)], 12). Because of this, they could only
be played on Pathé machines.
57   Adorno 1990 [1928], 5.

would not appear until the 1950s and the advent of stereo recording at the close of the decade).[58] Sound becomes reified: a reflection in the mirror. In the colour advertisement that Pathé placed in magazines, the same image is reused and elaborated upon. This time the flapper is blonde-haired; in keeping with the fashion of the era, her lips are painted bright red; she wears a fluid, silk-like blue gown, a pearl choker necklace, and a bright, narrow bracelet. The illustration of the youth holding the gramophone record is framed by a mirror-like border: an image within an image.

A third – more complex – rendition of Pathé's 'Mirror of the Voice' trope is to be found in a 1929 series of photographs by French photographer François Kollar. Kollar was commissioned by Pathé to produce a series of portraits of French theatre and cinema actress Marie Bell, capturing her reflection as it appeared in a perfectly reflective mirror-record (as seen on the cover of this book). The smooth surface (or 'skin') of the disc is a promise of sonic perfection, and we may feel compelled to contrast the neatly defined features of Marie Bell with the faceless (and nameless) female workforce toiling in pressing plants to produce such 'perfect surfaces' (see Chapter 2). Kollar's 'phono-vision' series presented a persuasive, eerily glamourous photographic translation of the firm's 'mirror of the voice' motto – as captured and composed by the male gaze. The disc literally becomes a black mirror, with the reflective potentials of the shellac substance fully realised.

In its series of advertisement, Pathé strategically played on the well-known reflective properties of shellac (discussed in Chapter 1). It is worth noting that there exists a material, historical link between gramophone records and looking glasses. As early as 1596, for instance, Dutch merchant and historian John Huygen van Linschoten had written about the mirror-like shine of shellac. Long before Berliner's made it a crucial material of sound, shellac was most readily associated with the realm of vision and reflectivity. As noted in Chapter 1, shellac moulded handheld mirrors had been popular artefacts in late Victorian households: the substance was mobilised in the manufacture of both records and mirrors (it serves to coat the backs of mirrors, or to frame them). The now-forgotten discourse around the reflective properties of records had a long history and continued to be palpable through the first three decades of the twentieth century. It must

---

58   For a technological and sociocultural history of the development of high-fidelity sound, stereo-recording and audiophilia, see Barry (2010). Though the term high fidelity became widely used in the late 1950s, the recording industry had been concerned with sound fidelity since its early days. In a 1940 manual, 'fidelity' was defined as 'faithfulness of sound reproduction' (Anon. [1940], 119).

be noted that at the time gramophone records could still be experienced in conjunction with other shellac-based domestic objects, most of which were typically associated with femininity (including hand-held mirrors, photographic cases and hairbrushes).

In the 1910s and 1920s, Edison-Bell's 'Velvet Face' records, distributed in Europe, also exploited this long cultural association: the smoothness of the recorded surface was presented as a promise of acoustic perfection. In German, the term 'Spiegel' ('mirror') also served to casually designate the 'side' of the record on which the label was glued. It may be remembered that, until the International Zonophone Company and Odeon introduced the first double-sided records (respectively in 1903 and 1904) in Germany, records were one-sided.[59] Unsurprisingly, Odeon's first advertisement for double-sided records (which were sold across Europe) used the visual trope of the mirror to display the two sides of the disc, playing on the material and symbolic polysemy of shellac discs. The advertising cartoon showed an anonymous, stretched out hand, firmly holding a disc in front of a mirror.[60] The magnified 'face' of the disc filled up the looking glass – substituting itself to the human face one would have typically expected there.

Significantly, the immaterial 'smoothness' of sound was associated with – and conditioned by – the actual haptic 'smoothness' of the lacquered disc: the eye, moving over the surface of the artefact, could therefore anticipate what the ear may hear. 'Brilliance' and 'brightness' were two terms historically related to sound fidelity. The brilliance – a term which contemporary listeners still use in a metaphorical way – was first and foremost understood as a visual phenomenon. Discourses on the 'brightness' and 'brilliance' of recorded sound – predicated upon the *visual* scanning of sound grooves – significantly multiplied in the 1920s and 1930s, most notably amongst the nascent community of home recordists (which recorded themselves on cellulose-coated records rather than shellac ones). In order to quickly check the quality of their recordings ('cuts'), early amateur recordists did not need to play them back. Rather, they looked at them, using a method known as the 'reflection method'.[61] By observing how the rays of light bounced off the cut surface of the record, they were thus able to 'readily determine the perfection (smoothness) of both walls of the groove'[62] and literally 'pre-view'

59   Ward (1990), 141; Block and Glasmeier (2018 [1989]), 264.
60   Chew (1981), 32.
61   Anon. (1940), 49.
62   Ibid.

the quality of the recording. In the reflection method, 'light falling on a perfectly smooth surface' would be 'reflected by that surface so that the angle of incidence is equal to the angle of reflection': the phenomenon was known as that of specular reflection. The surface of a well-cut record 'behave[d] like a mirror'.[63] On the other hand, 'a record appear[ing] dull or gray' would be marred by a high level of 'surface noise' and was therefore discarded with no need to play it back.[64] In record-pressing plants, the same visual method would be applied to quickly monitor the acoustic quality of commercial pressings. Amateurs also tested the speed of their record players visually by building 'remarkably accurate stroboscopic speed tester[s] on a piece of cardboard'.[65]

The visual and tactile methodologies developed to test recordings also help us understand that the term 'brilliant', which came to be used to describe 'sound that is good in respect to reproduction of high frequencies',[66] initially referred to the visual appearance of the disc. Perfect sonic reproduction was closely linked to visual perfection: reproduction and reflection coincided. In the late 1930s, the New York-based company Audio Devices perfected smooth, ungrooved records ('Audiodiscs') aimed at the niche yet growing market self-recordists. The blanks fashioned by the firm were described using the flattering, and familiar, analogy of the mirror: 'Audiodiscs are modern instantaneous recording discs. So perfect is the surface of an Audiodisc that it has the appearance of a darkened plate-glass mirror'.[67] However, beyond the actual reflectivity of discs and formal resemblance between sound-recording, reflecting and mirroring, a more complex process of (self-)identification may take place between the listener and the record. As evidenced by the advertisements described at the beginning of this section, Pathé's commercial strategy largely relied on a psychological mechanism of identification or self-objectification (and self-dramatisation). The fashionable youth depicted on Pathé record sleeves could either be the recording diva, or the record-buying listener – more likely, the two women merged into one figure, or one objective image. Similarly, the human subject and its mechanical reflection, the voice and the mirror, coalesce. In a narcissistic paroxysm, what seems to matter is less the recorded content than the captured image of the listener as she identifies with the recording, sinking into it as it were. The

63 Masschelein-Kleiner (1995), 17.
64 Anon. (1940), 49.
65 Wilson and Webb (1929), 249.
66 Anon. (1940), 116.
67 Ibid., 36.

'virtual photograph' represents the most idealised – or distorted – version of the self: so that the record becomes a deforming surface of self-projection, the site of a privatised spectacle.

The idea of the record as a family photograph – or a mundane mnemonic device – owed notably to the fact that both phonography and photography relied on the 'plate' (*Platte*), a term which, in German, also meant 'record'.[68] Adorno pursued the theme in his 1934 essay, developing the comparison with mordant irony: 'records are possessed like photographs; the nineteenth century had good reasons for coming up with phonograph record albums alongside photographic and postage-stamp albums, all of them herbaria of artificial life that are present in the smallest space and ready to conjure up every recollection that would otherwise be mercilessly shredded between the haste and hum-drum of private life'.[69] If records are possessed, listeners are also literally possessed or contained by recordings. For Adorno, one may own desultory 'scraps' of oneself in the form of recordings, though the brief moment of possession cannot be equated to genuine self-knowledge – or access to musical culture. In other words, generic self-projection hinders or defers potential introspection, and ultimately constitutes an obstacle to knowing oneself (in the Socratic sense of the phrase) – taken further, it may even threaten the development of one's humanity. This is because, for Adorno, the mirror cannot be traversed: it becomes an opaque, distracting screen, severing the dialectical relationship that one may form (through the mediation of the object) with one's subjectivity. Adorno's writings anticipate Heidegger's critique of modern technology as 'a metaphysics that estranges people from truth and revelation'[70] – as formalised in the 'Question Concerning Technology' (1954).

The recording, for Adorno (and, later, Guy Debord), should therefore be disparaged as a parody of life, a deceiving simulacrum of experience. Adorno's moralistic thesis is connected to broader concerns regarding the perceived impoverishment of lived experience (its 'artificiality') in the modern metropolis. Berlin life certainly offered individuals an unprecedentedly artificial and mediated form of existence – with the city presenting itself as a proto, microcosmic 'society of spectacle' (to reuse Debord's terminology). At the time Adorno was writing, Berlin was a centre of musical consumption, trade and production: almost all record companies – including Lindström and the German Gramophone Company – were based there, making it the

---

68   Levin (1990), 32.
69   Adorno (1990), 58.
70   Poschardt (1998), 356.

German 'capital' of the recording industry.[71] Lindström owned five labels (Electrola, Odeon, Parlophone, Gloria and Homocord), 'account[ing] for a third of the turnover from German record sales' in 1928.[72] In Adorno's agile, trenchant writings, the gramophone appears both as a pretext and as an emblematic machine through which the state of the modern psyche and its many ideological ramifications can be articulated and minutely excavated.

In a provocatively poetic piece entitled 'La Bouche Cornue du Phonographe' ('The Phonograph's Horned Mouth'), first published in French in 1988, Charles Grivel further theorised the body of the machine as it progressively became indiscernible from the body of the listener. The piece, published exactly fifty years after Adorno's claims that records presented the listener with a 'sounding image' of herself, suggested that: 'The self exists as long as it can be shown to me. In full, on the surface, sound-matter, flux, substance. From inside out, then – inversely – from inside in: the road back is reproduction; I displace myself in returning. I come back'.[73] Accordingly, the self is not to be found in inner subjective culture (as proposed by Simmel) but much rather it is produced at the surface, realised on the plane of appearances – where the listener can visually apprehend (though not necessarily comprehend) herself. The surface reflects the objective culture rather than incorporating or absorbing it: it retrospectively produces the 'inside' in so much as it reflects the 'outside'.

In Grivel's phenomenological argument, the reproduction machine does not capture the pre-existing subject but '[gives] the subject an origin, making it symbolically first, automatically engendering ego to ego'.[74] The reproduction machine is first and foremost a dynamic *production* machine. What this further means is that there is no subject – or no subjective essence: rather, 'the subject arises from the object' (as argued by Serres) and, more radically, the object establishes the very possibility of selfhood – listening retrospectively creates a founding (yet fragile) ground for the emergence of identity. For Grivel, the event of listening further constitutes a moment of 'projection, but also retroprojection', akin to a mystical moment of phantasmagorical self-aggrandising.[75] Listening is inseparably connected to seeing – and, particularly, to seeing a world beyond the world: for, as Grivel continues, 'The machine-mirror reflects the *unheard-of* [...] even the

71 Gauss (2014 [2014]), 79.
72 Schrader and Schebera (1988), 117.
73 Grivel (1988) in Kahn and Whitehead (1994 [1992]), 34.
74 Ibid.
75 Ibid., 35.

unimaginable, the hidden within, the possible within'.[76] Though he does not
elaborate upon it, Grivel's text is implicitly indebted to the Lacanian theory
of the mirror stage (developed in the 1930s) where the self is identified – or
founded – in the dramatic moment of specular recognition, appearing in
and as an 'exteriority'.[77] In other words, the self is simultaneously constituted in the moment of mediation and of severance (or disconnection): for
the external image is accepted as a separate object. Theologian Thomas
Aquinas had already mused on the close interconnectedness of speculation
and mediation in the formation of knowledge, outlining the foundational
antecedence of mediation in perception: 'to see something in a mirror' meant
'to seize a cause by its effect'.[78] Lacan privileges an eminently visual model
for the formation of subjectivity, a bias which has since been questioned
and nuanced by further generations of psychoanalysts.

In her study on artificiality and the feminine voice in French nineteenth-
century narratives, Felicia Miller Frank notably retraces how the Lacanian
model was unsettled by alternative psychoanalytical models positing an
auditory mirror stage preceding the visual mirror stage. She draws most notably from Didier Anzieu, and his incisive interpretation of the metonymical
figures of Echo and Narcissus – personifying, respectively, sound and vision.
In the Greek myth, as retold by Ovid in his *Metamorphoses*, the chattering
nymph Echo, before being reduced to a ghostly reverberation, would 'always
answer back'.[79] She uses speech as a weapon, intentionally detaining Juno
'when she could have caught the nymphs lying with her Jupiter on the
mountainside'.[80] Juno (seeking revenge) condemns Echo to repeat 'the last
words' spoken by her interlocutor. Because Echo's handicap prevents her
from speaking first, of her own accord, she is doomed 'to wait for sounds
which she might re-echo with her own voice'.[81] One day Echo sees Narcissus
wandering through the woods, and falls in love with the indifferent, haughty
boy who, though he continually excites the nymphs' passions, only ever
'play[s] with [their] affections'.[82] Echo follows him from afar, waiting for an
opportunity to express her love. One day Echo finally obtains an opportunity
to 'speak' to the one she loves, only to awaken Narcissus's wrath with her
fragmentary stutter. The humiliated Echo withdraws and slowly wastes

76 Ibid., 58.
77 Lacan (2006), 76.
78 Quoted in Melchior-Bonnet (1994), 124; my translation.
79 Ovid (1974), 83.
80 Ibid.
81 Ibid., 84.
82 Ibid.

away, her body progressively disappearing, so that only her voice remains, 'anticipating the disembodied voice of a gramophone recording'.[83] Later on, Nemesis punishes Narcissus for his indifference. Returning thirsty from a hunt, Narcissus is made to lean over a spring to drink and, upon seeing his reflection, hopelessly falls in love with it. Narcissus, grown weaker with passion, finally perishes. In Anzieu's analysis, 'The legend well indicates the precedence the sound mirror has over the visual mirror, as well as the primarily feminine character of the voice and the connection between the emission of sound and the demand for love'.[84]

While she does not mention Anzieu, Barbara Engh relied on psycho-analytical models to understand 'the technologically disembodied voice', likening early phonographic listening to an 'acoustic mirror stage'.[85] Through recording technology, the voice became a discrete external object, enabling the listener to experience or apprehend herself as another, and more specifi-cally as an objectified other. Engh described the 'crisis' or 'techno-trauma' (to reuse Wolfgang Ernst's expression) which took place when the subject recognised her own voice on record for the first time. For her, the experience bore the same violence and ego-formative power as the one described in Lacan's mirror stage because it posited a radical frontier and rupture between the 'inside' and the 'outside' – while at the same time indicating that the inside was inexorably constituted *from the outside*. The moment of recording sensorially and cognitively dis-located the subject whilst simultaneously reassembling (or objectifying) her in the (apparently) homogeneous surface of the record. Recorded sound selectively repeats a trace of the actual, yet disappeared, body. Accordingly, the advent of phonography – understood as the image of sound – can be read as a variation upon the Echo and Narcissus motif: it offers a story where sound and vision – resonance and reflection – are bound together by symmetrical yet irreconcilable fates. This is notably because acoustic and visual spaces drastically differ: while sound unfolds in time, the image has an 'apparent immediacy'.[86]

This section ends with an evocation of Jean Vigo's poetic film *L'Atalante* (1934), which provides another, different entry point into the cultural imagination of recorded sound as a mirror. In one of the early scenes of the film, we see Juliette, a young country girl gazing steadily into a river to try and divine the face of her future lover, enacting an old superstitious

83   Ernst (2016), 59.
84   Quoted in Miller Frank (1995), 35.
85   Engh (1999), 55.
86   Ernst (2016), 58.

ritual (where water is fantasised as a divinatory mirror). The face of Jean, her husband-to-be, briefly appears on the shiny surface of the water. When Juliette eventually meets Jean – who is a sailor – she follows him to live with him on a barge. One night, as the boat stops in Paris, Juliette (who has never seen the capital city) impulsively leaves her companion; we see her wandering through the animated streets of the capital, caught in a sort of trance or childish fascination with the urban spectacle. The film ends with Juliette standing in a gramophone parlour by the Seine, listening to mariners' and rivermen's songs; trying to revive sonically, through recorded music, her connection with her lost lover. The journey of Juliette is a compressed journey through the ambiguous surfaces and states of modernity, from the country to the city, from seeing the future to hearing back the past – from the prophetic medium of water to that of the air which carries back the melodies of the past, and a memory of love. We move from the realm of hope and pre-vision to the realm of playback, nostalgia and repetition; we witness Juliette blooming from her archaic, peasant girlhood into modern wired womanhood.

Beyond the mirror: Film, photography and liquid records

The above section has retraced (and furthered) Adorno's discussion of the 'mirror function' of recordings, exploring the interplay between sound reproduction, self-identification and narcissism. Yet the mirror-record did more than providing a limited or alienating surface of reflection – and was not uniquely a sum of ephemeral, passing effects. Adorno's discourse on technological alienation must be contrasted with practices seeking to emancipate everyday media from its conventional field of application. In the interwar, recorded sound became a potent source of inspiration for experimental visual artists who were interested in the productive and creative potentials of the phonographic technology, envisioning it as an intermediary between sound and vision. Europe-based or Europe-born artists including László Moholy-Nagy, Man Ray, Germaine Dulac but also Marcel Duchamp, all experimented with the visuality of gramophone records, reinterpreting them in their own idioms. In doing so, they strove to emotionally reclaim and defetishise the recorded commodity.

Some artists (including Moholy-Nagy, Man Ray and François Kollar) closely photographed the surfaces of mass-produced discs, transforming them into exotic, unfamiliar objects and inviting viewers to see differently. It can be suggested that they questioned the trope of the perfect mirror by conceiving the record not as a surface of passive reflection or reproduction but as one of production – capable of suggesting alternative images of the real. Moholy-Nagy's 1927 close-up photograph of a gramophone record (entitled *Grammophonplatte*), with its grooves clearly delineated and visible, worked as an invitation to haptically (and not simply visually) interact with the medium of the disc: to immediately touch it, scratch it, deface it in order to 'offer without new instruments and without orchestra a *new way* of generating sound'.[87] Through photography, he further sought to merge temporality (*movement*) with spatiality (*image*), to reconcile Echo and Narcissus. The subtle mobility of sound was evoked, without being destroyed, by the static image: Moholy-Nagy's photograph made visible the actual topography – or landscape – of the record, magnifying its 'lands' and its 'grooves', as they were poetically known as the time.

The topography of the record was also rearranged, this time in a three-dimensional manner, by Marcel Duchamp's in the series of *Rotoreliefs* (*Optical Discs*) he first produced in Paris in 1935.[88] The set of six discs was made out of cardboard, and every disc featured a colour lithography on each of its sides, becoming animated at the speed of 33rpm. The rotoreliefs were to be played on a wall-hanging unit or 'revolving magnetized turntable'.[89] The lithographs showed a series of stylised circular objects, including eclipses, lanterns, and hoops which came to life when they were spun. The installation – which could be described as a form of primitive cinema apparatus – offered a technologically updated version of the magic lantern shows of the late seventeenth century.

Duchamp was not alone in trying to articulate the sonic within the visual realm. A few years before him, in her short film *Disque 957* (1928/29), French experimental filmmaker Germaine Dulac attempted to capture 'visual impressions' of two Chopin preludes. Dulac's abstract cinema can be situated within the lineage of the pattern films of the late 1920s, perhaps best exemplified by Dudley Murphy and Fernand Léger's *Le Ballet Mécanique* (1924). The latter proposed a fragmented cinema of lights, ruptures, abrupt alterations, appearance and disappearance – simulating the fractured temporalities and sensory overload characteristic of modern urban experience. *Disque*

87 Moholy-Nagy (1969 [1925]), 31. This anticipated the record-based experiments conducted by Milan Knížák or Christian Marclay, and many other sound artists and turntablists, including, in the 1990s, the German techno DJ Thomas Brinkmann who would directly etch and modify the surface of his vinyl records in order to produce new sound patterns.

88 Four more editions of the *Rotoreliefs* would be issued in the following thirty years, respectively in New York (1953; 1963), Paris (1959), and Milan (1965). See Block and Glasmeier 2018 [1989], 126.

89 Block and Glasmeier (2018 [1989]), 126.

957, the first of a series of three shorts dedicated to the idea of visualising sound, could be understood as an animated version of Moholy-Nagy's still photograph *Grammophonplatte* (1927). Whilst Moholy-Nagy grappled with the challenging technical issue of spatialising (sonic) time, Dulac used the dynamic film as a means to understand the very nature of movement and audio-visual synchronicity. Her film begins with a close-up of a disc spinning and multiplying itself on the screen, progressively dissolving into a close-up vision of hands swiftly stroking piano keys. The images of the spinning disc and the hands of the pianist are superimposed – suggesting a coincidence between recorded sound and live music, object and performance. *Disque 957* – which partially restores human tactility to the realm of the mechanical, revealing the fingers of the player – can be contrasted with Man Ray's haunting photograph *Gramophone et main en bois* ('Gramophone and Wooden Hand') (c. 1930), made at the same period in Paris. The photograph shows an articulated manikin hand steadily grasping the pickup arm of a gramophone. A simple reading of the photograph may suggest that contact with the inanimate body of the gramophone has turned the living hand into a petrified piece of wood. Yet a paradoxical intimacy – and a fulgurant impression of tenderness – also arises from the picture: the machine and the artificial hand seem to gently communicate through touch. The wooden hand almost appears to be more sensitive, with its hyperbolised joints and finely veined surface, than a human hand would be (the natural acoustic properties and high tactility of wood must be recalled here). Man Ray's photograph also invites us to reinterpret Dulac's film: the focus on the hands of the pianist, it may ultimately be suggested, expresses less their human qualities than their mechanical state: for the hands – hyperbolised and made autonomous by the artifice of cinematic framing – move in an oblivious, trained fashion, suggestive of an automaton.[90] Eventually, what is staged by the two artworks is a form of contagion or cross-sensibilisation between the natural and the artificial as well as the suggestion that one doesn't necessary have to 'decide' between the two.

Dulac defined 'avant-garde' as 'any film whose technique, employed with a view to a renewed expressiveness of image and sound, breaks with established traditions to search out, in the strictly visual and auditory realm, new emotional chords [...] The sincere avant-garde film has this fundamental quality of containing, behind a sometimes inaccessible surface, the seeds of the discoveries which are capable of advancing film towards

---

90   A parallel could be drawn here to the pianola's hybrid status; Hannah Lewis (2020) has notably analysed the visual and narrative ubiquity of mechanical pianos in 1930s French cinema.

the cinematic form of the future'.[91] She thus posited a relative autonomy or self-sufficiency of cinema: for the technological eye of the camera may see more (and better) than the human eye. According to her, the act of seeing may renew and revitalise the act of hearing, and reciprocally: we may further add that it is because of their radical difference that the two senses may be able to stimulate one another.

Dulac's and Man Ray's works can be discussed in relation to the aspirations and experimentations of the international 'New Photography' movement, investigating the affinities between man and machine, organic and artificial life, nature and industry. The New Photography's major achievements were displayed in the international 'Film und Foto' exhibition held in Stuttgart in 1929. Of endless – and foremost – interest for the New Photographers was the objective world, with its composite textures, materialities, surfaces, and reflections. As neatly summarised by the exhibition's director Gustaf Stotz: 'We see things differently now, without painterly intent in the impressionistic sense. Today things are important that earlier were hardly noticed: for example, shoe laces, gutters, spools of thread, fabrics, machines, etc. They interest us for their material substance, for the simple quality of the thing-in-itself; they interest us as a means of creating space-form on surfaces, as the bearers of the darkness and the light'.[92] And the thing-itself – with its dizzying, fluctuant iridescences and multifaceted aspects – mirrored 'the self of modern life [...] built up out of tiny bits of rubbish, of hurtful glances, of bumps on busy streets, of signs glimpsed out of the corner of the eye'[93] – a fragmented self which in turn acquired an ambiguous thing-like quality.

The New Photographers' concern with material life – and the (photo-) sensitivity of objects – aligned with some of the cinematic production of the pre-Hitler period, as meticulously analysed by left-wing theorist Siegfried Kracauer. For Kracauer, objects came to play an increasingly strong part in the dramatic action of film in the first part of the 1920s, notably in the dark, silent cinema of Carl Mayer. Until then, objects had served the comic tradition (or slapstick): but Mayer lent a dramatic significance to 'details of locomotives, wheels, telegraph wires, signal bells'.[94] These immediately modern motifs – which also appeared in the New Photographers' works – did not constitute an indifferent backdrop, extraneous to the action: rather, they influenced and conducted the drama to its inexorable denouement,

---

91   Quoted in Zurbrugg (1999), 98.
92   Cited in Turner and Badger (1988), 1977.
93   Leslie (2016), 171.
94   Kracauer (1966 [1947]), 103.

allowing Mayer to develop a new model of cinematic narration – which was to influence a new generation of filmmakers. As he fully gave voice to mute objects, Mayer reciprocally 'dragged' bodies 'to the realm of objects', detachedly filming them as fragments, so closely observed that they lost their familiar – and eventually always relative and precarious – human appearance.[95] In other words, the frontiers between 'human' and 'nonhuman' were indefatigably questioned, dissolved, and ultimately displaced, by the camera eye. Benjamin forcefully showed, for instance, how the movie-camera (and the technique of close-up) allowed for human beings to see for the first time quotidian yet hitherto unknown bodily gestures. Photography and cinematography could thus allow individuals to perceive the profoundly gritty, grainy surfaces of everyday life (though they would equally reveal its confounding beauty). Blown out of proportion, objects and bodies became unfamiliar and potentially frightening, shocking or awe-inspiring – a form of mutual arising and enriching was sought between humans and nonhumans, on the absolute condition that audiences were 'receptive to the times in which they live[d]'.[96]

Benjamin's argument of the 'optical unconscious' expanded upon the thesis succinctly put forward by Moholy-Nagy in his 1925 theoretical manifesto *Painting, Photography, Film*: 'Art attempts to establish far-reaching *new relations* between the known and the as yet unknown optical, acoustical, and other functional phenomena so that these are absorbed in increasingly abundance by the functional apparatus'.[97] The photographer would further his argument in a 1932 essay entitled 'A New Instrument of Vision', listing the passive and active uses of photography, and reaffirming its power to forcefully reveal previously unforeseen (and imperceptible) aspects of life.[98] In other words, photography *augments* and *transforms* life: it follows that a photography which only reproduced or repeated 'existing relationships' between objects was, for Moholy-Nagy and the Bauhaus School, sterile.[99]

Even though the 'primitive idea of the equivalence of graphic and musical creation could only come into being at a time when the crudity of the product corresponded to the crudity of the hardware',[100] Moholy-Nagy's theses articulated an alternative to monolithic critiques – or glorifications – of technology. The work of Moholy-Nagy and his students at the Bauhaus was

95   Ibid.
96   Moholy-Nagy (1969 [1925]), 43.
97   Ibid., 30; emphasis in the original.
98   Clarke (1997), 192.
99   Moholy-Nagy (1969 [1925]), 30.
100 Poschardt (1998), 349.

guided by the idea of ceaseless movement, translation and estrangement. Through visual effects and technical manipulation (double-exposure, high contrast, collage, montage, close-up), the framed fixity of images was to be overcome, producing an effect of unlimited motion. But these experiments did more than revealing the optical unconscious: they encouraged audiences to conceive of, and eventually interact with, the outer world in a different way. As well as generating *'new, previously unknown relationships'*,[101] the technology potentially held a transformative – and therefore revolutionary – power over the life it sought to capture.

Technological ambivalence

The Bauhaus embraced material culture politically, in the largest sense of the word; the material, sensory mode of knowledge-making it promoted can be compared to the approach pioneered in Soviet Russia by artist, architect and stage-designer Vladimir Tatlin, an important theorist of materiality. In 1923, Vladimir Tatlin was appointed as the head of the 'Section for Material Culture'[102] at the Petrograd State Institute for Artistic Culture. In his enthused introductory address, he defined material culture as: '1) Research into material as the shaping principle of culture; 2) research into everyday life as a certain form of material culture; and 3) the synthetic forming of new material and, as a result of such formation, the construction of standards for new experience'.[103] Ultimately, Tatlin believed that a new everyday, and as such a new man, could be fashioned through research into alternative materialities.

During the revolution of 1917, Tatlin had enthusiastically pleaded for the destruction of the old regime. His was a highly ideological, utopian and politicised version of material culture as that which, literally, could alter the medium – and therefore the meaning – of life itself. Because Revolutionary Russia had abruptly broken free from its historical mould and traditions it needed to urgently invent a common myth of the future: Tatlin's half-finished *Monument for the Third International* (1919–1925), on which he was working at the time he began running the 'Section for Material Culture', embodied such hopeful longings for a novel future. His monument took the form of a tower made of steel, glass and wood and intended as a response to the

---

101 Moholy-Nagy (1969 [1925]), 30, emphasis in the original.
102 A position he would hold between 1923 and 1925.
103 Quoted in Lehman (2015), 181.

'bourgeois' Eiffel Tower (1889). The tower, Tatlin believed, would serve as a vast conference space for the Communist Party in Petrograd. It was never completed: the tantalisingly unfinished (and potently un-functional) construction clearly makes visible the parallels between utopian ideology and material research.

The applied research conducted at the Bauhaus and the Petrograd State Institute for Artistic Culture was underpinned by a strong theoretical and poetic drive: indeed, practice was intimately woven with theory (one could not exist without the other in the general project to 'change life'). In addition to artists' and practitioners, thinkers of the interwar period (those of the Frankfurt School especially) also recognised that polyvalent (and turbulent) technological objects were part of – and partially organised and produced, at their own speed and scale – a larger composite socio-historical order, arguably contributing to fashioning a novel sensibility and 'innervation' of society as a whole. They wrote about the contemporary technological condition in nuanced, plural ways – as a fragmented age of machines rather than a total 'machine age'. Although they were not the first to interrogate technologies (a preoccupation which can be traced back to Plato's *Timaeus*), they coined nuanced responses to it.

Rather than identifying, in a deterministic way, the spiritual alteration machines might perversely induce in the national psyche, theorists of the Frankfurt School – anticipating the work of Marshall McLuhan – also explored the sensory, affective and somatic disruptions (as well as extensions) that new technology may bring to individual (yet interconnected) sentient bodies. While they remained acutely aware of the dangers of mass-manipulation, their writings differed from the discourses on technology emerging in the US in the previous century where thinkers and poets such as Ralph Waldo Emerson, Henry David Thoreau, or Walt Whitman had discussed the relationship between technological advances and the making of a new national, progressive republican identity. What US thinkers underlined was the emancipatory potentials of machines as well as their monstrosity. Emerson, in particular, grew increasingly disillusioned with the machines he had admired as a young man, suggesting that they diminished and impoverished human capacity for expressive freedom, imagination and even emotion.[104] In a characteristically bold aphorism, he warns us that 'the machine unmans the user'[105] – a sombre and techno-deterministic

---

104 Kasson (1999 [1976]),127.
105 Quoted in Kasson (1999 [1976]), 124.

prognosis which would fleetingly reappear in the twentieth century media writings of Friedrich Kittler.

The debate on technology and democratic identity was still a significant component of early twentieth-century critical thought – though it assumed different forms on the continent. European thinkers of the interwar, rather than questioning the already ordinary technological large-scale innovations, directed their attentions towards discrete, pervasive media objects – observing smaller, and often domestic, technological devices which Mumford associated with the 'neotechnic' phase of technology (relying on 'alloys, synthetics, electronics, and automation'[106]). With Benjamin and others, the often monolithic or deterministic discussion on 'technology' became an open discussion on (mixed) media materialities and hybrid, techno-human sensibilities. Despite many theoretical and methodological divergences, a recurrent feature of interwar thinking about materiality (as explored by artists, designers, and theorists) was that it provided understandings of matter as a dynamic and potentially 'active' source – rekindling premodern discourses on the agentive power and influence of things.[107] Benjamin was especially sensitive to the instability, ambivalence and potential reversibility of cultural production, noting how 'documents of culture' were inseparably tied to barbarism and exploitation.[108]

The short story 'The Street of Crocodiles', penned by Polish writer and illustrator Bruno Schulz in 1934, offers us a complementary reflection on the felt ambivalence of matter – as a mediator between the physical and the psychic realms. Schulz's protagonist is a textile merchant who strives to make a Golem-like dummy out of leftover pieces of fabric, sewing together 'the thousand scraps, the frivolous and fickle trimmings'.[109] The author depicts his childlike, immensely physical 'love of matter as such, for its fluffiness or porosity, for its unique mystical consistency'.[110] The short story delivers a tender, yet gravely metaphysical, reading of matter as a secular 'key' (or medium) to access the spiritual world. Conversely, if matter is active and sensitive, its ill-use and mishandling may constitute a source of danger. Schulz's narrator warns us that '[m]atter never makes jokes: it is always full of the tragically serious. Who dares to think that you can play with matter, that you can shape it for a joke, that the joke will not be built in, will not eat

106 Miller Frank (1995), 164.
107 Böhme (2014), 3.
108 Benjamin (2003), 392.
109 Schulz (1963), 45.
110 Ibid., 52.

into it like fate, like destiny?'.[111] Bearing in mind Schulz's poignant words, the next chapter explicitly returns to the cultural specificity of shellac in order to expose its material and ideological toxicity in the context of the two World Wars.

Bibliography

Adorno, Theodor W. 1990. 'The Curves of the Needle'. Translated by Thomas Y. Levin. *October* 55: 48–55.

Adorno, Theodor W. 1990. 'The Form of the Phonograph Record'. Translated by Thomas Y. Levin. *October* 55: 56–61.

Anon. 1940. *How to Make Good Recordings*. New York: Audio Devices, Inc.

Benjamin, Walter. 2003. On the Concept of History. In *Selected Writings, vol. 4, 1938–1940*, eds. Howard Eiland and Michael W. Jennings, 389–400. Cambridge, Massachusetts: Belknap Press / Harvard University Press.

Berenbaum, May. 1995. *Bugs in the System: Insects and their Impact on Human Affairs*. Cambridge, Massachusetts: Perseus Books Blake.

Block, Ursula and Michael Glasmeier. 2018 (1989). *Broken Music: Artists' Recordworks*. New York: Primary Information.

Böhme, Hartmut. 2014. *Fetishism and Culture: A Different Theory of Modernity*. Translated by Anna Galt. Berlin and Boston: De Gruyter.

Boym, Svetlana. 1994. *Common Places: Mythologies of Everyday Life in Russia*. Cambridge, Massachusetts and London, England: Harvard University Press.

Chew, Victor K. 1981. *Talking Machines*. London: Science Museum.

Citton, Yves. *Médiarchie*. 2017. Paris: Editions du Seuil.

Clarke, Graham. 1997. *The Photograph*. Oxford, New York: Oxford University Press.

Derrida, Jacques. 2010. *Copy, Archive, Signature: A Conversation on Photography*. Stanford, California: Stanford University Press.

Engh, Barbara. 1999. 'After "His Master's Voice"'. *New Formations* 38: 54–63.

Ernst, Wolfgang. 2016. *Sonic Time Machines: Explicit Sound, Sirenic Voices, and Implicit Sonicity*. Amsterdam: Amsterdam University Press.

Föllmer, Moritz. 2013. *Individuality and Modernity in Berlin*. Cambridge: Cambridge University Press.

Gauss, Stefan. 2007 (2014). 'Listening to the Horn: On the Cultural History of the Phonograph and the Gramophone'. In *Sounds of Modern History: Auditory Cultures in 19th- and 20th-Century Europe*, ed. Daniel Morat, 71–100. New York, Oxford: Berghahn Books.

111 Ibid., 54.

Gilbert, Marianne. 2017. 'Plastics Materials: Introduction and Historical Develop-
ment'. In *Brydson's Plastics Materials. Eighth Edition*, ed. Marianne Gilbert, 1–18.
Amsterdam, Boston, Heidelberg, London, New York, Oxford, Paris, San Diego,
San Francisco, Singapore, Sydney, Tokyo: Elsevier.

Grivel, Charles. 1994 (1992). 'La bouche cornue du phonographe', translated by
Douglas Kahn and Gregory Whitehead. In *Wireless Imagination: Sound, Radio and
the Avant-Garde*, ed. Douglas Kahn and Gregory Whitehead, 51–75. Cambridge,
Massachusetts and London, England: The MIT-Press.

Gronow, Jukka. 2003. *Caviar with Champagne: Common Luxury and the Ideals of
the Good Life in Stalin's Russia*. Oxford: Berg. Ebook.

Gronow, Pekka and Ilpo Saunio. 1999. *An International History of the Recording
Industry*. London and New York: Cassell.

Hailey, Christopher. 1994. 'Rethinking Sound: Music and Radio in Weimar Germany'.
In *Music and Performance during the Weimar Republic*, ed. Bryan Gilliam, 13–36.
Cambridge, New York, Melbourne: Cambridge University Press.

Hake, Sabine. 1994. 'Urban Spectacle in Walter Ruttmann's *Berlin, Symphony of
the Big City*'. In *Dancing on the Volcano: Essays on the Culture of the Weimar
Republic*, ed. Thomas W. Kniesche and Stephen Brockmann, 127–142. Columbia:
Camden House.

Isherwood, Christopher. 1998 (1939). *Goodbye to Berlin*. London: Vintage.

Kaes, Anton, Martin Jay, Edward Dimendberg, eds. 1995. *The Weimar Republic
Sourcebook*. Berkeley, Los Angeles, London: University of California Press.

Kasson, John F. 1999 (1976). *Civilizing the Machine: Technology and Republican
Values in America, 1776–1900*. New York: Hill and Wang.

Kracauer, Siegfried. 1966 (1947). *From Caligari to Hitler: A Psychological History of
the German Film*. New York: Princeton University Press.

Lacan, Jacques. 2006. Ecrits: *The First Complete Edition in English*. Translated by
Bruce Fink. New York and London: W. W. Norton & Company.

Lehmann, Ulrich. 2015. 'Material Culture and Materialism: The French Revolution
in Wallpaper'. In *Writing Material Culture History*, eds. Anne Gerritsen and
Giorgio Riello, 173–190. London, New Delhi, New York, Sidney: Bloomsbury.

LeMahieu, Daniel. 1982. The Gramophone: Recorded Music and the Cultivated
Mind in Britain between the Wars. *Technology and Culture* 23 (3): 372–339.

Leslie, Esther. 2016. *Liquid Crystals*. London: Reaktion.

Levin, Thomas Y. 1990. 'For the Record: Adorno on Music in the Age of Its Technological Reproducibility'. *October* 55: 23–47.

Lewis, Hannah. 2015. '"The Music Has Something to Say": The Musical Revisions of
L'Atalante (1934)'. *Journal of the American Musicological Society* 68 (3): 559–603.

Lewis, Hannah. 2020. 'The *piano mécanique* in 1930s French Cinema'. *French Screen
Studies* 20 (3–4): 158–179.

Maisonneuve, Sophie. 2002. 'La constitution d'une culture et d'une écoute musicale nouvelles: Le disque et ses sociabilités comme agents de changement culturel dans les années 1920 et 1930 en Grande-Bretagne'. *Revue de Musicologie* 88 (1): 43–66.

Maisonneuve, Sophie. 2006. 'De la machine parlante au disque: Une innovation technique, commerciale et culturelle'. *Vingtième Siècle. Revue d'histoire* 92 (4): 17–31.

Masschelein-Kleiner, Liliane. 1995. *Ancient Binding Media, Varnishes and Adhesives.* Translated by Janet Bridgland, Sue Walston and A. E. Werner. Rome: ICCROM.

Melchior-Bonnet, Sabine. 1994. *Histoire du miroir.* Paris: Hachette.

Miller Frank, Felicia. 1995. *The Mechanical Song: Women, Voice, and the Artificial in Nineteenth-Century French Narrative.* Stanford, California: Stanford University Press.

Miller, Henry. 2009 (1936). *Black Spring.* Richmond: Oneword Classics.

Moholy-Nagy, László. 1969 (1925). *Painting, Photography, Film.* Translated by Janet Seligman. London: Lund Humphries.

Naylor, Gillian. 1968. *The Bauhaus.* London: Studio Vista.

Nora, Pierre. 1989. 'Between Memory and History: Les lieux de mémoire'. *Representations* 26: 7–24.

Ospina-Romero, Sergio. 2019. 'Ghosts in the Machine and Other Tales around a "Marvelous Invention": Player Pianos in Latin America in the Early Twentieth Century'. *Journal of the American Musicological Society* 72 (1): 1–42.

Ovid. 1974. *Metamorphoses.* Translated by Mary M. Innes. London: Penguin Books.

Poschardt, Ulf. 1998. *DJ Culture.* Translated by Shaun Whiteside. London: Quartet Books.

Roth, Joseph. 2003. *What I Saw: Reports from Berlin 1920–33.* Translated by Michael Hofmann. London: Granta Books.

Rothenbuhler, Eric W. and John Durham Peters. 1997. 'Defining Phonography: An Experiment in Theory'. *The Musical Quarterly* 81 (2): 242–264.

Roy, Elodie A. 2017. 'Broken Records from Berlin: The Place of Listening in *People on Sunday* (dir. Curt and Robert Siodmak/Edgar G. Ulmer, 1929)'. *Sound Studies* 3 (1): 33–48.

Schrader, Bärbel and Jürgen Schebera. 1988. *The "Golden" Twenties: Art and Literature in the Weimar Republic.* New Haven and London: Yale University Press.

Schulz, Bruno. 1963. *Cinnamon Shops and Other Stories.* Translated by Celina Wieniewska. London: Macgibbon & Kee.

Silva, João. 2019. 'Portugal, Mechanised Entertainment, and the Second Industrial Revolution'. In *Music and the Second Industrial Revolution*, ed. Massimiliano Sala, 57–80. Turnhout: Brepols.

Sterne, Jonathan. 2003. *The Audible Past: Cultural Origins of Sound Reproduction.* Durham and London: Duke University Press.

Taussig, Michael. 1993. *Mimesis and Alterity: A Particular History of the Senses.* New York, London: Routledge.

Turner, Peter and Gerry Badger. 1988. *Photo Texts.* London: Travelling Light.

Ward, Alan. 1990. *A Manual of Sound Archive Administration.* Aldershot: Gower.

Wilson, Percy and George W. Webb. 1929. *Modern Gramophones and Electrical Reproducers.* London, Toronto, Melbourne and Sydney: Cassell and Company.

Zielinski, Siegfried. 2018. 'A Media-Archaeological Postscript to the Translation of Ernst Kapp's *Elements of a Philosophy of Technology* (1877)'. In *Elements of a Philosophy of Technology: On the Evolutionary History of Culture,* by Ernst Kapp, Translated by Lauren K. Wolfe, 251–266. Minneapolis and London: University of Minnesota Press.

Zurbrugg, Nicholas. 1999. 'Marinetti, Chopin, Stelarc and the Auratic Intensities of the Postmodern Techno-Body'. *Body and Society* 5 (2–3): 93–115.

4. Detonations: Shellac at war

Abstract:

Chapter 4 attends to the toxic transformation of shellac in the two World Wars, when it became a key substance in the manufacture of detonating compositions, hand grenades, and bombs and was rationed by governments (thus curtailing record-making operations and intensifying research into substitutes, including PVC). Drawing from Catherine Malabou's radical theses on 'destructive plasticity', it theorises the material and ideological instability of shellac as well as its recycling, exploring the dominant discourses associated with recorded sound. In particular, it draws attention to the trope of phonographic listening as a means to repair both individual and social bodies broken down by war, showing how this discourse was recuperated by governmental bodies.

Keywords: war, phonography, plasticity, toxicity, vinyl, shellac

During the two World Wars, shellac was governmentally rationed, stockpiled and repurposed as a strategical material by the War Authorities, serving as 'a protective or waterproofing varnish on the exterior or interior of numerous stores; as an adhesive on paper and textile discs in fuses, shells, etc., and for the attachment of labels on packages; as a binding agent for certain detonating compositions and for ignition compositions in pyrotechnic stores; as an ingredient of special varnish for marking TNT exploders, and of Kieselguhr varnish for coating the inside of pyrotechnic cylinders; and as a means of securing roller pins and pallet fans in watches'.[1] Drawing from archival sources and secondary literature, this chapter examines the liquidation of shellac during the two World Wars when global supply chains as well as patterns of record manufacturing were disrupted, altered and progressively dismantled. Shellac shortages were contributing factors to the research into alternative synthetic materials such as vinylite (which

---

1   Parry (1935), 172–173.

Roy, E.A., *Shellac in Visual and Sonic Culture: Unsettled Matter.* Amsterdam: Amsterdam University Press, 2023

DOI 10.5117/9789463729543_CH04

became the basis for long-playing records). As it was progressively integrated within the wider ecology of war, shellac – a nontoxic substance – became a hazardous material by association. In the meantime, phonographic listening was also instrumentalised and ideologically repurposed both on the home front and in conflict zones – while the war generated novel and ambiguous sites for recording experimentations (notably in the context of German prisoner-of-war camps).

This chapter offers a reflection on the physical and symbolic indeterminacy of media materials – underscoring the fact that their trajectories cannot be contained, predicted or arrested. Materials may 'pass around some obstacles, dissolve some others and bore or soak their way through others still'.[2] In other words, they fluctuate, and may generate and enter into new compositions, both on a discrete and more global level. What makes shellac so thought-provoking to study is its mobile, in-between status, its flexibility and restlessness. What makes it so significant is, paradoxically, its indeterminacy where the finished or 'perfect' artefact (such as the gramophone disc) never completely conceals, dominates or petrifies the material. This in turn may call for a widened material ontology of what media objects are and what they might do.

In recent years, several media and material culture studies have surveyed the lingering toxicity of materials, investigating well-known contemporary harmful synthetics such as PVC but also historical materials of culture.[3] These studies seek to assess the impact of chemical substances on environments as well as on the individuals who handle and consume them. For instance, fashion historian Allison Matthews David has extensively researched the hazardous material culture of dress and ornament, notably identifying the poisonous (and, in some instances, deadly) chemicals involved in the production of dyes and cosmetics (including lead and arsenic) from the late nineteenth to the early twentieth century.[4] Measuring the full implications of the expression 'fashion victims', she has shown how the cultural formation and valuing of physical beauty, and notably the discursive pairing of morals and aesthetics in the nineteenth century, also contributed to the ruining and (self-)exploitation of the female body.

Toxicity, however, is not merely an attribute of matter. Contamination does not occur unidirectionally but can be understood as a process of mutual transformation. Innocuous, nontoxic materials such as shellac

2   Bauman (2000), 2.
3   Maxwell and Miller (2012), 59; Devine (2019a).
4   Matthews David (2015).

may become harmful in combination with other substances, or in certain usages. As noted before, shellac is an adhesive in more than one sense, and on a number of scales. On a discrete level, it may bind several chemical or material components together. On a macroscopic level, the shellac trade 'connected' an extended range of epistemes, institutions, industries and locales across the globe, while practices of record-listening partly authorised or intensified fresh forms of social and musical conviviality.[5]

And while it may operatively bind heterogenous elements together, shellac is also that which isolates and keeps materials apart (with its insulating properties), resonating with Samuel Weber's definition of a medium as that which 'makes the connection possible as division'.[6] The material ambivalence of shellac anecdotally and discretely relates to broader processes of alteration, modulation and transposition – where 'gramophone tunes' may intimately and uncannily converse with the detonations of bombs. The emphasis on materiality and sites of transformation allows us to concretely explore the moments when 'culture' and 'barbarism' indissolubly entwine in a relation of active contemporaneity and cooperation. There may be a literal 'explosive' or self-destructive capacity of shellac as it becomes something else entirely.

It is important to return here to the lesser-known side of 'plasticity'. Etymologically, the term both signals the capacity of matter to receive form – to be moulded or imprinted – and its active capacity to annihilate form by reassembling itself in a radically novel (or alien) configuration. In French the term 'plastique', as well as denoting malleability, refers to 'an explosive substance made of nitroglycerine and nitrocellulose, capable of causing violent explosions'.[7] The writings of Catherine Malabou have theorised the radical implications of plasticity, considering complete deformation as the most extreme instance of metamorphosis. Weaving together philosophy and neurobiology, Malabou's reflection addresses in particular the shattering event of 'destructive plasticity' in individuals affected by traumatic and post-traumatic disorders (as well as by degenerative neurological pathologies such as Alzheimer's) – or in those whose facial features have been irretrievably and ruinously 'erased' (through causes including ageing and illness). Plasticity is theorised as a desertion or a total 'flight of identity', where nothing remains of a former self.[8]

5   Maisonneuve (2006).
6   Quoted in Ernst (2013), 105.
7   Quoted in Mawani (2015), 164.
8   Malabou (2012), 12.

While transformation (or metamorphosis) may constitute 'a form of redemption, a strange salvation',[9] as well as a potentially fecund moment of development, Malabou insists on 'destructive plasticity' as a radical, violent and irreversible alteration by which a former self gets entirely annihilated. In what could be described as a moment of auto-alienation (or auto-alteration), individuals '[become] new people, others, re-engendered, belonging to a different species. Exactly as if they [have] had an accident'.[10] Here, the term 'elasticity' – which implies that a return to a former form is possible – is opposed to plasticity – where no return may ever be achieved. For Malabou, the principle of destructivity 'hides within the very constitution of identity':[11] even unactualised, plasticity therefore signals a virtual (negative) horizon of identity. It is inseparably bound with futurity.

Although Malabou primarily stresses the sphere of individual subjectivity, her writings further invite us to relate the concept of plasticity (understood as pure potentiality) to a reading of history. An understanding of historical becoming as plastic is one which embraces the catastrophe – or accident – to be. In that sense, Malabou's reflection provides an important underwriting and critique of the archaeological method (as notably deployed in Freudian psychoanalysis). No return is possible. Her philosophical argument points to the impossibility of haunting (or self-haunting): for radical disappearance leaves no trace and no surplus, therefore offering no hope of redemption, no future in the past, no messianic possibility. Plasticity corresponds to a moment of total liquidation and total forgetting (a form of death in life). There is literally no ground left to dig up, no debris to tentatively grasp, no step to retrace: identity has become non-identical to itself. Accordingly, reconstruction cannot occur either: the site of annihilation can only be marked by a silence. Malabou's concept of plasticity, opposing Derrida's spectrality and Benjamin's topographical reading of history, points to the staged and strenuous character of any 'return'. Where (messianic) hauntological understandings insist on cyclicality – and processes of reparation –, 'destructive plasticity' exposes the violent irreversibility of individual (and historical) becoming. The following only partly engages with the specific instances (and radical implications) of destructive plasticity so subtly delineated by Malabou. However, it retains something of her suggestive conceptualisation of 'plasticity' as a moment of crisis or irrevocability – beyond the redemptive and positive model offered by theories of recyclability and circularity. By

9   Ibid.
10   Ibid., 13.
11   Ibid., 37.

focusing on shellac during the two World Wars, the chapter emphasis moments of crises and radical transformation – taking us beyond the familiar understandings of phonographic cultures.

First World War: Mechanical bodies and phonographic data

The First World War corresponded to a period of restructuring for the gramophone industry during which its supply chains and modes of production had to be rearranged. In August 1908, two months after its pressing plant was launched, the Hayes record factory had produced 28,000 thousand discs monthly;[12] by 1914 it held the national monopoly for the manufacture of records, and was able to 'supply all the records for the British market'.[13] In the UK, the passing of the Defence of the Realm (Consolidation) Act in November 1914 granted the government power to take over factories. Within a few weeks of the declaration of war in August 1914, the Hayes record-pressing plant was retooled to produce munitions whilst record-making operations and recording session would continue to take place.[14] Shellac was mostly used on the outside of munitions, for 'A shell dipped in a solution of shellac – which dries instantly – [made] the shell more effective'. In addition to this, hand grenades were sealed and waterproofed by shellacking them, according to a technique perfected one century before by English landscape painter and inventor Joshua Shaw.[15] Because of its insulating properties, shellac was also used to coat the inside of shell casings to prevent dangerous fillings from leaking and 'reacting with the metals of which the containers [were] made'.[16] The British Ministry of Munitions needed a stable source of shellac and in late 1916, to circumvent 'excessive speculations' in Calcutta, a direct agreement was concluded with British India to obtain shellac at preferential rates.

It must be noted here that India had traditionally played a central role in the history of explosives and their dissemination across the Western world; in the seventeenth century, the East India Company had imported

12   Jones (1985), 85.
13   Ibid., 89. The Gramophone Company further controlled 'a wide manufacturing and distribution network in Europe and British India', owning five factories in Europe, including one in Austria-Hungary (Aussig) and two in France (Chatou, Ivry). See Jones (1985), 86–87.
14   Blake (2004), 25.
15   In 1816, exiled in Philadelphia, Shaw experimented with percussion caps for guns: he sealed and waterproofed the whole unit by shellacking it. Brown (1999), 174.
16   Ibid., 150.

saltpetre in large quantities to manufacture gunpowder, a mixture which Alfred Nobel once described as 'possess[ing] a truly admirable elasticity which permits its adaptation to purposes of the most varied nature'[17] – its blasting properties making it a key ally in mining and civil engineering.[18] From 1917 onwards, shellac – which could also be described as an elastic substance – was centralised and stockpiled for governmental use. It was also resold (in rationed quantities) to individual customers (including Allies and national businesses).[19] Four governmental stores were constituted in the country, situated in greater London and in the Liverpool area (these were in Southall, Nine Elms, Crewe and Birkenhead): in order to store the material as efficiently as possible, the British Government sought advice from the experienced managers of the Gramophone Company. By May 1917, a total of 130 tonnes of shellac were stockpiled in Britain. Amongst the government's main clients were the French Government, Nobel Explosives, and the Hayes record-pressing plant, which had also secured an important war material contract.

Alfred Clark, the managing director of the Gramophone Company, explained that he decided to retool the factory in order to keep his staff together. In the course of the Great War, the number of employees at Hayes was multiplied by four and the profits of the factory dramatically increased.[20] Ninety per cent of munition workers were women[21] who were involved in the manufacture of a variety of time-fuses, shell cases and aircraft parts.[22] The munition workers of the Amatol section of the plant, filling shells on site, were not protected from the powders. Lyddite, which was at the time the main military explosive in Britain, discoloured and jaundiced their skin – so that the Hayes shell-filling workers were familiarly known as the 'canaries'. TNT progressively replaced Lyddite in the last years of the war, leading to the insidious poisoning and disfiguration of the women and girl who were exposed to it.[23]

With its strategic and massive use across the modern weapon industry, shellac became part of lethal and toxic assemblages, whilst record-listening

17 Quoted in Brown (1999), 80.
18 The states of Bihar and Bengal were amongst the largest suppliers of gunpowder, and remained so until the early nineteenth century when newly discovered deposits of caliche were discovered in Chile. See Brown (1999), 13.
19 TNA, MUN 4/24/54.
20 Martland (1997), 73.
21 Ballin (1996), 238.
22 Lowe, Miller and Boar (1982), 105.
23 Brown (1999), 159.

was at the same time promoted as a unifying or humanising pastime. Despite shellac shortages, new records never ceased to be pressed for the home market and the British troops:[24] the Gramophone Company recorded and commercialised over 1,000 new titles[25] in the course of the Great War. During the war, the introduction of suitcase gramophones (notably the Decca Dulcephone 'trench' model) notably marked the beginning of a more mobile culture of record-listening. Gramophones could now be found on ships, in hospital wards, in war factories (to stimulate production), in army barracks and even in trenches, amongst other sites. One of the dominant assumptions concerning the values of phonographic listening regarded its perceived 'therapeutic' and socially regulatory functions (a common place repetitively used by the British government). The belief in 'music's healing properties' was not new in itself: it could be traced back to the third book of Plato's *Republic* and had been particularly prevalent in Western music theory in the late 1700s and 1800s, exploring the 'notion that music possesses intrinsic qualities generating clearly definable emotional and mental effects'.[26] However, in the wartime context, the Platonic intuition was rationalised and systemised to serve alleged therapeutic ends. The British War Office issued a number of notices celebrating the comfort brought by recorded sound, encouraging civilians to donate their old records – as well as listening devices – to soldiers, including those in hospitals. This was tied in with an enduring – and deterministic – belief in the therapeutic and humanising powers of music, conceived of as a means to bring together a dispersed or shattered community.

In some instances, gramophones appeared to have been used as quick substitutes for human warmth and companionship – and could be compared to ambiguous mothering (or nursing) machines. In a poem entitled 'In a Soldiers' Hospital II: Gramophone Tunes', written shortly after the war, British nurse Eva Dobell fondly reminisced over the socialising and revitalising function of the gramophone in the military hospital. She described how playing the device together metonymically reanimated or re-formed, part after part, the weakened bodies of crippled soldiers. The 'Welsh boy', lying in bed with a lame leg, 'wind[s] the handle half the day'; his one-armed friend 'picks out the records he must play'; another one 'beats the time' with his crutches. A deaf, shell-shocked gunner 'listens with puzzled patient,

---

24   As it was forbidden for manufacturers (and dealers) to export records outside the UK, except to British troops.

25   These included popular patriotic numbers, military records, descriptive records, spoken words, classical music, and so on. See Roy (2018), 29.

26   Fauser (2013), 127–128.

smile' while the rest of the company 'join in from their beds, / And send the chorus rolling round'.[27] The idiosyncratically broken bodies of the soldiers are replaced with – and surpassed by – a collective human automaton[28] in a moment when 'flesh was turned into machine as machine was turned into flesh'.[29] What emerges is a conception of a distributed automated body not dissimilar from the individual yet dispersed machinic body later visualised by Umbo in his *Roving Reporter* collage. Human limbs become technological appendices (or 'extensions').

Interestingly, Dobell's poem is chiefly preoccupied with the bodies of soldiers, stressing visible scars and external signs of suffering: nothing is said about the mental traumas they may suffer. In Dobell's text, the belief in music's healing power was crudely shifted from the sphere of the mind (and the soul) which had so concerned Plato to be given a very literal, materialistic sense: the gramophone instrumentally mended broken bodies rather than attending to invisibly broken spirits, and it did so by transforming individuals into an efficiently synchronised and automatised collective being. The records, uncannily echoing the properties of shellac, were imagined as a quick social adhesive to 'repair' the social. Dobell's (hyper-deterministic) glorification of the machine's power to assemble people (in the industrial, assembly-line sense of the term) is not an isolated occurrence.

The literature of the period (especially soldiers' correspondence) bears frequent, unexamined claims celebrating the sociability of the gramophone, suggesting that listening to recorded music, for those who had access to the technology, had become a widespread recreational pastime or, even, addiction – comparable, in certain instances, to a sonic opium. British soldiers' letters indicate that gramophone discs were sometimes played to soothe the nerves 'during the tense evenings' before attacks.[30] Paul Klee, who spent part of the war stationed in Munich as an infantry reservist, suffered with increasing distress the sound of the gramophone 'plaguing the barracks again', malignantly encroaching upon his rare moments of solitude. In his wartime diaries the painter disparaged the machine as a 'piece of hell always near [him]', and described the 'heads' of his comrades 'grin[ning] around it' like 'devilish masks'[31] – a passage magnifying, in a caricatural way, the scene of collective mirth remembered by Eva Dobell. In a 1916 letter, a

27   Dobell (2006 [2004]), 208.
28   Roy (2018), 34.
29   Bruno (2007 [2002]), 148.
30   Brittain (1986 [1933]), 286.
31   Klee (1965), 392.

Figure 4: 'Le gramophone' ('The gramophone'), wood engraving by Jean-Émile Laboureur, c. 1918–1921. Bibliothèque nationale de France.

comfortably-off British lieutenant named Walter A. MacClean, interned in the German POW camp of Crefeld, wrote to a music firm in Birmingham urgently requesting that 'a quantity of Gramophone Records' were sent to him in order to entertain him and his companions: as well as a sedative, recorded sound was hailed as a stimulant. At the time, exports of gramophone were strictly forbidden – except to British troops.[32] A wood engraving made shortly after the war by French artist Jean-Émile Laboureur expresses the sense of easy conviviality and blissful oblivion associated with record-listening: a

32   Letter to Scotcher & Sons Ltd, dated May 1916. TNA, FO 383/192

small party of riders is gathered around a suitcase gramophone, drinking and smoking, their bodies pleasurably relaxed (see Figure 4).

The celebration of recreational phonography must be contrasted with the recording experiments which took place in German and Austrian prisoner-of-war camps. During the war, a total of two and a half million soldiers were captured and interned in Germany.[33] In late 1914, 'hundreds of thousands of soldiers of non-European origin' – notably those from the French and British colonies – began filling up the POW camps. About three hundred kilometres south of Crefeld lay the military town of Wünsdorf and its Halfmoon Camp, where thousands of African and Indian soldiers were held in captivity. These camps provided the basis for large-scale data-gathering projects as well as the place where the 'relatively new technologies of photography, film, and phonographic recording could be implemented on a mass-scale'.[34] Nearly three thousand recordings of prisoners' voices, dialects and songs were made at Halfmoon Camp, where the Royal Prussian Phonographic Commission had set up a recording studio.[35] The commission, established in 1915 with funds from the Kaiser, was headed by psychologist and musicologist Carl Stumpf, and came under the technical and logistical direction of linguist and English teacher Wilhelm Doegen. It mainly counted amongst its ranks philologists, as well as a smaller contingent of anthropologists, whose avowed goal was to further develop, through sound-recording technology, the late nineteenth-century colonial project of sampling, measuring and controlling otherness.[36] The commission aimed, in Doegen's words, to 'systematically fix the languages, musics and sounds of all tribes residing in German POW camps on sound discs in combination with corresponding texts, according to methodological principles'.[37] Amongst the recordings made at Halfmoon Camp,[38] more than four hundred of them were of displaced Indian soldiers. The collection of colonial records, later to be housed at the *Berlin Lautarchiv* ('Archive of Voices'), points to several levels of captivity: the colonial subject was captured and entombed twice over, being a 'socially dead' prisoner as well as a captive of the record, literally *buried alive*.[39] One may remember here the original 'entombment' of the lac insect in the resin, and the labour concealed in the production of shellac. Speeches and more occasional folk

---

33 Scheer (2010), 287.
34 Ibid., 279.
35 Lange (2015).
36 Stumpf was the commission's sole musicologist. See Scheer (2010), 302.
37 Doegen (1925) in Lange (2015), 88.
38 Other recordings were made at other camps throughout the war.
39 Lange (2015), 90.

songs and instrumentals were recorded onto wax discs from which a metal negative was made, allowing in turn to make multiple copies in shellac. In what appears to be a last, ironic twist, the voice of the displaced soldier unknowingly encountered the displaced recording resin, the primary yet invisible 'substance' of the record.[40]

Contrary to the easily transportable cylinder phonograph, which allowed for sounds to be played back immediately after being recorded and, for this reason, had been the technology of choice in early ethnographic fieldworks (such as the ones carried out by the Berlin Phonogramm-Archiv), disc-recording technology was bulky and more awkward to operate.[41] In addition to this, it implied a phase of 'post-production' comparable to that of photographic processing. The moment of playback was deferred rather than immediate: the sounds engraved on wax discs were unplayable until they were 'remediated'. A metal negative had to be made from the wax disc, which in turn allowed to make a playable shellac disc. Shellac discs were (for purposes of memorialising) superior to wax cylinders in that, contrary to the latter, they offered a greater archival stability. Furthermore, additional copies could always be made from the metal matrix (wax cylinders deteriorated with each occurrence of playback). Because of the promises of durability and reproducibility it offered, gramophone recording technology was used in POW camps: the device didn't have to be portable because prisoners were personally taken to the recording equipment (rather than it being brought to them). The fact that the recording apparatus was permanently set up also allowed for a rationalistic, optimal studio environment to be devised. During a recording session, the performer would stand 'before the gramophone's horn, which was mounted into a partitioning board, behind which the technician operated the recording device itself, thus isolating the speaker from the recording equipment'.[42] No contact or interaction occurred between the prisoner and the recording operator.

The anthropologists and other scientists of the Phonographic Commission construed the camp as 'an enormous archive of sounds waiting to be transferred onto media where they could be preserved and reproduced at will, for study, teaching, or entertainment' – as well as 'the perfect setting

---

40   It must be noted that many international recording campaigns continued to take place during the war, notably with Fred Gaisberg's expeditions for the Gramophone Company. In 1917, he travelled to Milan to record 'war songs' and visited various prison camps with his recording equipment. See Gaisberg (1948 [1947]), 75.

41   Scheer (2010), 293.

42   Ibid., 305–306.

for collecting sound recordings'.[43] In this context, the recording technology became rationalised as a data-producing and data-gathering machine. The mass of prisoners constituted a readily accessible and plunderable reservoir of sonic data: recording was an extractive process aligning with a broader colonial ideology which conceived of the other as available. In 1906, Carl Stumpf – discussing the aims of the Berlin Phonogramm-Archiv – had already insisted on the significance of sound archives as 'a necessary corollary of our colonial aspirations in the highest sense'.[44] The Berlin Phonogramm-Archiv, emerging from the nascent procedures of comparative musicology (which derived its methods from social evolutionary theory), notably ambitioned to make evolution – and otherness – 'audible'.

The large-scale programme carried out within the POW camps was to further these ambitions. The extracted voice of the prisoner represented an essentialised specimen or sample in at least two ways – being both a 'sample' of the physiological body and of the 'racial' body. On the one hand, each recording was accompanied by a file identifying the prisoner and containing the text he had been 'asked to read out in his own dialect or language' as well as a phonetic transcription of it and, in some cases, a German translation of it.[45] On the other hand, identification was suspended: the recording process could be read as an operation of depersonalisation where what was prioritised was the extraction of a fetishised essence – to be rehoused in the body of the *Lautarchiv* (as repository of anonymous voices). The depersonalisation is notably evidenced by the fact that prisoners were frequently asked to read aloud exactly the same script in order to facilitate the work of linguistic comparison.[46]

This anonymising or flattening of voices – which were only valued as re-ified data or proofs of difference – can be read in contrast with the process of subjectification or narcissistic identification evoked in Chapter 3 in relation to phonographic 'mirrors'. Prisoners never listened to their own recordings, notably because immediate playback was a technical impossibility. In order to listen to them, researchers themselves had to wait until the recordings had been processed and transferred to shellac: but at this point, recorded voices were once more remediated and translated back into textuality, their sonicity – with its particular, irreducible inflections – repressed and rationalised. The fact that phonetic transcriptions were systematically made

43 Scheer (2010), 279–280.
44 Quoted in Radano and Olaniyan (2016), 10.
45 Scheer (2010), 306.
46 Ibid.

strongly suggests that the disc as artefact – as well as any possibility of ethical listening it may authorise – were deemed to be of secondary importance.[47]

The scientific recording sessions conducted at the Halfmoon Camp resonate with the commercial wartime phonographic experiments carried out by US recording engineer Fred Gaisberg for the Gramophone Company. Gaisberg notably travelled to Milan in 1917 in order to record war songs and establish 'a local pressing and matrix-pressing plant'.[48] In Italian prison camps, he recorded deserters' folk songs, dances, and spoken words; a pressing plant was hastily fashioned 'out of parts scavenged from [...] junk heaps' since 'new machinery could only be head for armaments'.[49] Shortly before the end of the conflict, in the first days of October 1918, Gaisberg's brother Will was sent over to France by the Gramophone Company with his equipment: with the help of an assistant, he recorded the Royal Garrison Artillery 'firing a barrage of gas shells as a prelude to the British troops' advance upon Lille'.[50]

While the formal experiments carried out in German POW camps had sought to scientifically measure and assess the 'essence' of otherness by making it audible, Gaisberg and his assistant set out to capture the 'essence' of the conflict. The recording was commercially released by the Gramophone Company's HMV trademark in the months following the Armistice under the generic title 'Gas Shell Bombardment' (HMV 09308). Although the authenticity of the recording should be questioned (some effects were certainly contrived in the studio to heighten the dramatic element), the small eight-inch gramophone disc, advertised by the Company as 'a historical record which should be in every home',[51] constituted a sonic postcard from the battlefront. It appeared as an exemplary instance of domestication and 'trivialisation': the recent war was miniaturised 'so that it would become commonplace instead of awesome and frightening'.[52] Incidentally, the material on which the detonations of shells were reproduced was also the material which had (partially) been used in their manufacture, producing a literally ob-scene encounter. Although this passed unremarked, the kinship between the phonographic and the weapon industry was anecdotally and uncannily manifested in the release of the 'Gas Shell Bombardment' disc.

47   The documentary film *The Halfmoon Files* (dir. Philip Scheffner, 2007) aimed to 'repatriate' the voices to their speakers, and recontextualised the sound archive and its foundations within the larger colonial history.

48   Gaisberg (1948 [1947]), 74.

49   Ibid.

50   Blake (2004), 30; see also Rust (1975), n.p. and Lowe, Miller and Boar (1982), 105.

51   Rust (1975), n.p.

52   Mosse (1991 [1990]), 126.

Second World War: Shortage and recycling

The First World War marked a temporary slowing down in the global circulation of shellac. The trade was energetically resumed and intensified during
the interwar period (see Chapter 3). The Second World War clearly revealed
its vulnerability when shellac resources became increasingly inaccessible to
European and North American countries (notably due to Japanese military
operations in South Asia in late 1941, disrupting access to the material). In the
US, shellac stocks were placed under the control of the War Production Board
in April 1942;[53] in Britain, they were similarly centralised and stockpiled
as they had been during the previous war. The uses of shellac during the
Second World War multiplied, and plugs for shells and bombs (known as
'transit plugs') were massively manufactured in Indian shellac factories and
shipped to Britain. Over three million of these plugs were manufactured
between 1941 and 1945 at the Angelo Brothers factory,[54] while jutlac – a
shellac-jute laminate used to replace rationed critical materials such as tin
plate and plywood – was also developed there. The making of records was
deemed of very secondary importance.

During the Second World War, the record industry – which had consider-
ably expanded during the interwar – faced a number of shortages: raw
materials such as steel, aluminium and paper were all rationed, prompting
gramophone plants to settle for less satisfactory substitutes. For instance,
the wax which was used to varnish and protect recording blanks originated
from Russia, and could no longer be procured.[55] The shellac shortage was
more critical as it directly concerned 'a vital constituent of the 78rpm',
effectively 'restrict[ing] the manufacture of the discs themselves' as no
efficient substitute existed.[56] In Britain, from August 1942 onwards,
a number of appeals were made in the press (including in specialised
magazine such as *The Gramophone*) to invite civilians to donate their
unwanted records so that they could be melted and transformed into
new ones – notably to entertain meritorious war workers (in factories
and at the front).

The national salvage campaign launched in 1942 by the British (now
Royal British) Legion was advertised through explicative posters displayed

53   Broven (2009), 14.
54   Anon. (1956), 49.
55   Paper was also rationed, affecting the quality and appearance of magazines such as *The Gramophone* which had to reduce its size. Pollard (1998), 61; 63.
56   Ibid., 61.

in record dealers' windows as well as pamphlets. On the one hand, the British Legion insisted on the 'official purposes' of recorded sound, extolling the virtuous impact that light music programmes such as the BBC's 'Music While Your Work' had on factory workers – allegedly boosting both morale and productivity. Music was thus considered to be of greater value in the factory than it was at home, so that domestic ownership and private enjoyment of records and music-playing devices (while not strictly antipatriotic) nonetheless appeared to be detrimental to the collective war effort. On the other hand, the Legion also highlighted that shellac should be reserved for the higher purposes of war and, accordingly, not be wasted to make gramophone discs: it follows that surplus discs – a form of musical waste – were recycled both materially and symbolically. However, not every record would be indiscriminately accepted: musical waste itself was hierarchised and organised according to a specific scale of desirability. Records deemed materially inferior – those using a low percentage of shellac for instance – were excluded from the salvage drive and only discs issued by the biggest and most 'reliable' recording firms were accepted (including HMV, Columbia, Parlophone, Decca, Zonophone and Brunswick records).[57] A promotional event was organised outside the Legion's main office in London to reinforce the message of the posters.

An archival photograph shows the mediatic personalities of the day – including the comedienne Nellie Wallace, the comedian Sandy Powell and the singer Anne Shelton – duly queuing up under a British flag to donate their unwanted gramophone discs. The staged photograph echoed with the propagandist photographs of long recruitment queues circulated by the British Government during the two wars to incite young men to join the army. In February 1943, the Gramophone Company (EMI) and Decca issued a joint statement regarding the preoccupying shellac situation. The appeal, printed in *The Gramophone*, appealed to record collectors' patriotic feelings, sententiously informing them that:

> The further maintenance of adequate record supplies will depend upon the goodwill and readiness of the public to return old and unwanted records, because only by this means will manufacture continued. We ask users not only to give up their old records but to encourage their friends who may no longer be interested, but who may have old records, to give them up also.

57   Blake (2004), 144.

Some millions of scrap records are needed. In the last ten years over one
hundred million records have been sold and it should be possible for the
quantity required to be returned by the public. It does not matter what
condition the records are in provided they are not broken. All of them
will be reground and aid in making new record material.[58]

The allusion to the fact that the donated records should not be broken is
intriguingly paradoxical: because the discs were meant to be reground,
it seems to be of minor importance whether they were physically intact
or not. On a technical level, it may be suggested that the machinery used
for grinding or cleaning records only accepted whole discs, or that labels
needed to be visible and legible to assess the material composition of the
discs (and check that they came from companies deemed acceptable).[59] We
may also suggest that Decca and EMI would not actually destroy all records
and that a portion of them was redistributed or resold. It is unclear how
efficient the above appeal was.

As *Gramophone* magazine founder and columnist Compton Mackenzie
noted the following year, parting with one's lovingly assembled record
collection constituted a form of renunciation, a 'sacrifice' which 'must
be made only if the individual is prompted by his own sense of duty'.[60]
The above statement, however, selectively left out any reference to the
sentimental value of sonic artefacts, emphasising as it did the functional
dimension of the record as a (raw) commodity. Though there is no evidence that similar collecting campaigns took place on continental Europe
(where shellac usage also came under governmental control), the British
shellac salvage drive resonated with similar transatlantic programmes.
In the US, charitable organisations such as the Entertainments National
Service Association (ENSA) or the National Federation of Music Clubs
collected unwanted record-listening equipment, records, and needles to
be redistributed amongst servicemen – while the armed forces 'purchased
machines for their units so as to enable musical recreation among the
soldiers'.[61] In September 1942, just one month after the first shellac drive
was organised, the manufacture and supply of musical instruments (including gramophones) came under a governmental order: the Board of Trade
proposed that, owing to the shortage of raw materials, production may

58   Quoted in Pollard (1998), 69.
59   I am grateful to Ron Geesin for discussing this point with me.
60   Quoted in Pollard (1998), 69.
61   Fauser (2013), 119; see also Pollard (1998), 62.

even 'cease entirely in the near future'.[62] Although the manufacture of gramophones for civilian uses was effectively curtailed, production was not completely interrupted: it continued in order to supply Governmental departments, the British Red Cross and other charitable organisations with listening devices. In 1942, governmental organisations were asked to provide the Board of Trade with an estimate of their annual requirements of portable gramophones for the following year – as well as to indicate the purposes to which the devices would be put. The exchanges of letters between the Board of Trade and its correspondents revealed how covetable record players had become. The BBC answered that a production of 20,000 players per annum 'would meet [their] requirements', while emphasising that record playing facilities stimulated production in munition factories and, accordingly, shouldn't be dispensed with. A letter from the Royal Air Force similarly underlined that the need for gramophones '[couldn't] be exaggerated' as they provided entertainment and education – in the form of 'recorded classical music and language courses' – 'on isolated sites at which a great many personnel [were] stationed'. The production of gramophones was thus maintained throughout the war for the benefit of governmental and para-governmental activities.

V-Discs, vinyl and extended temporality

Although important stockpiles were accumulated both in the UK and the US, shellac became increasingly expensive and difficult to secure as the war progressed. The blockade of the Malayan Peninsula by the Japanese in late 1941 interrupted trade routes, making access to India almost impossible and prompting the US Office of War Information to fund research into synthetic substitutes.[63] Two articles published in April 1942 in the US trade magazine *Broadcasting* – the month the material was placed under the control of the War Production Board – surveyed the shellac situation and its implications for the record manufacturing industry, underlining the latter's almost total dependence on imports from British India and the uncertain future of shipping infrastructures.[64]

Research into synthetic plastics had already made substantial progress in Germany in the 1920s and 1930s as the country – economically marginalised

62   TNA, BT 64/1738
63   Read and Welch (1976), 424; see also Pollard (1998), 61.
64   Anon. (1942a), 10.

in the aftermath of the Great War – sought to gain material self-sufficiency.[65] In the context of wartime scarcity, research was energetically resumed in the US, receiving financial inputs from both the transcription industry and, more substantially, the government.[66] Vinylite, however, was initially developed not as a replacement for shellac itself but rather for the highly flammable nitrocellulose and acetates which surfaced radio transcription discs (or 'instantaneous' discs). In early 1941, those materials were strategically reserved for the manufacture of munitions.[67] Vinylite, itself a 'semi-byproduct of munitions materials', came from research into nitrocellulose substitutes: it was not restricted and could be obtained from firms such as the Union Carbide & Carbon Corporation based in West Virginia, the biggest and earliest manufacturer of PVC in the US (it had begun commercialising it in the early 1930s).[68] Although 80% of the Corporation's vinylite went into the production of war materials, enough was left to meet the needs of transcription companies.[69] Transcription records were critical to the broadcasting industry – serving for instance for 'news broadcasts', 'rebroadcast of speeches' or yet again 'special Government programs'.[70] Each year over one million blanks were consumed by transcription companies across the US: they couldn't be reused or recycled.[71]

Before it became a recording substrate, vinylite was first and foremost a 'surfacing' substance which was applied to a thicker basis: during the war, transcription blanks would have a solid glass base (as aluminium, which had been previously favoured, had ceased to be available for non-war uses). The idea that vinyl could be used to make 'full' gramophone discs was not initially welcomed by record companies. Interviewed in 1942, a Decca representative feared that 'discs made of this substance would not be desirable as they [were] too thin and flexible' while Frank Walker of RCA outrightly rejected the idea on the ground that vinyl records would be too 'terrifically expensive' to manufacture.[72] The doubts raised by both the RCA and Decca companies regarding the future of vinylite must be resituated

65   Westermann (2013).
66   It must be noted that in 1931 RCA had already experimentally introduced (as a novelty) an early form of vinyl named, after shellac, Vitrolac. Substitutes, however, were more expensive than shellac, which partially explains why vinyl wasn't widely used in record manufacturing prior to the Second World War. See Harris (2017), n.p.; Osborne (2012), 67.
67   Anon. (1942a), 53.
68   Osborne (2012), 67.
69   Anon. (1942a), 53.
70   Anon. (1942b), 10.
71   Ibid.
72   Ibid., 53.

within their particular histories and ambitions. It is worth observing that RCA had abortively and sporadically experimented with a vinyl precursor – a novelty substance named Vitrolac, after shellac – in the early 1930s, soon giving up its costly experiments.[73]

During the war, technicians at the Decca firm (under the guidance of Arthur Haddy) devoted much of their efforts to improve the audio quality of shellac discs. By 1945, the team had devised an audiophilic disc capable of offering a 'full-frequency range recording' covering 'almost the entire range of frequencies heard by the human ear': having finally and laboriously reached such high-fidelity results, it was therefore not in the interest of the company to relinquish shellac.[74] The development and mass introduction of vinyl-based records were therefore much more closely connected to governmental efforts than they were to the recording industry per se – especially because the latter's activities were severely curtailed by the recording ban imposed by the American Federation of Musicians in 1942 (the ban lasted until 1944). The US recording industry therefore lacked the material resources it required to continue its activities (access to shellac, repertoires or musicians was now suspended). In such a context, record companies' unsold gramophone stocks – which had been accumulating for years – benefitted from a new lease of life and were reinjected into the US market as 'new sensations' to meet audiences' undiminished appetite for recorded sound.[75] Reissues also gained in popularity.[76]

Union president James C. Petrillo's motivations for introducing the ban were diverse, though an important contention was that 'canned music' had displaced and deskilled live musicianship, depriving musicians of important sources of revenues (an argument which resonated with the discussions prevailing within the British Musicians' Union at the same time). It was further claimed that musicians were not fairly compensated when their discs were broadcast and played on jukeboxes.[77] The ban, however, 'concerned only commercial recordings for public broadcast'[78] and unionised members were allowed to record – albeit without payment – for government agencies such as the Office of War Information and the armed services – and to be involved in subtle forms of sonic propaganda. It follows that the first examples of discs making substantial use of vinylite were issued via the War

73   Harris (2017), n.p.; Osborne (2012), 67.
74   Osborne (2012), 67.
75   Roy (2021a), 214.
76   Blake (2004), 196.
77   Ibid., 195.
78   Fauser (2013), 79.

Department of the United States, which launched its V-Discs programme
(under the guidance of recording engineer Captain George R. Vincent) in
1943. The Office of War Information had 'needed immediately an unbreakable plastic-type material that could be easily shipped throughout the world
and on which propaganda programs could be recorded and distributed with
least possible delay'.[79] An atmosphere of emergency therefore underpinned
the adoption of the plastic disc.

The V-Discs destined to entertain US servicemen and women, distributed
on a strictly non-profit basis, were entirely made out of vinyl – though they
still played at 78rpm on standard equipment. The material endurance
of these '12-inch practically unbreakable platters', as a 1945 descriptive
brochure put it,[80] meant they could be dropped onto military ships and other
strategical sites with much reduced risks of breakage; their durability further
ensured the dissemination of their symbolic, morale-boosting contents.
The repertoire featured on V-Discs was varied, including contemporary
songs, dance bands, symphonic, classical and, to a smaller extent, patriotic
music. One of the features which made them different from commercial
recordings was that their musical contents were prefaced by 'a short, cheering
message spoken by one of the featured artists'.[81] Records would be shipped in
waterproof boxes containing twenty records each (and two hundred needles):
within two years, over three million V-Discs were liberally disseminated
around the world.[82] The unbreakable V-Disc appeared as an intermediate
cultural object between the gramophone disc and the long-playing record.
A narrower groove could be cut into vinyl, effectively resulting in a longer
playing time (one side would contain up to six minutes of sound). It may be
argued that the circumstances in which plastic records were first introduced
contributed to their early appeal and aura, where 'vinyl' became durably
conflated with 'victory'. Paradoxically, the sturdy, indestructible V-Discs
were initially conceived of as ephemeral sonic artefacts: all matrices and
surviving pressings were 'pledged to be destroyed' at the end of the war.
Nonetheless, though all metal masters were effectively smashed in 1949,
many pressings survived in private collections.[83]

In the wake of the non-commercial V-Discs programme, the recording
industry examined more attentively the lucrative possibilities afforded by

79   Read and Welch (1976), 424.
80   Quoted in Fauser (2013), 116.
81   Blake (2004), 196.
82   Fauser (2013), 116.
83   Blake (2004), 197.

vinyl; Columbia and RCA were the first two companies to develop vinyl-based commercial discs. The 33rpm long-playing record was officially unveiled by Edward Wallerstein (president of Columbia) in June 1948 while RCA introduced its competing 7-inch 45rpm record the following year. In the UK, the LP would be launched by the Decca company exactly two years after it was first commercialised in the US, and EMI released its first long-playing records – as well as the UK's first 45s – in 1952, at which point vinylite had displaced shellac as the primary ingredient in record-making.[84] It must be noted that already in the 1930s Columbia had issued, in limited numbers, long-playing records for specific use in cinemas. These records however, which were played 'from the inside to the outside at 33rpm', couldn't be used on standard domestic equipment and therefore never reached the broader public.[85] Efforts at perfecting the LP further suffered from 'the inadequacy of the materials available for the manufacture of the disc itself'.[86]

As indicated above, the novelty of PVC was that, in addition to its sturdiness, it allowed for the engraving of much finer grooves: manufacturers were therefore able to press a larger quantity of grooves into the disc. While 78rpm discs had contained an average of 85 grooves, 'EPs and LPs averaged between 224 and 260 grooves per inch'.[87] Vinyl records were pressed according to the same process as shellac records (see Chapter 2), although in the early days the pressing cycle for vinyl lasted six times longer[88] – and the surplus material couldn't be remelted. The later automated presses – such as the Lened automated record press patented by Flusfeder and Palmer in 1964 in New Jersey, and still in use in record plants around the world today – allowed to streamline and combine the operations. The Lened press was summarily described by Flusfeder's son in a spare, functional language reminiscent of the efficiency of the machine itself:

A record machine is really two machines, an extruder to squeeze the raw plastic pellets into a plug to which the press itself applies 1,800 pounds of pressure at 320 degrees Fahrenheit to inscribe the grooves. No concession is made to aesthetics. A record press as designed by my father is pure function. Button pressed: PVC plug moved to the places that contain the negative imprints of the master disc; plug flattened; labels applied;

---

84   Pollard (1998), 85–87; Katz (1985), 13.
85   Pollard (1998), 83.
86   Ibid.
87   Melillo (2020).
88   Affelder (1947), 235.

hole punched. After the disc is pressed, and cooled, it is shunted along
to the trimming-station part of the machine, where the excess vinyl is
sliced away.[89]

The homogenous and replicable composition of vinyl discs contributed
to drastically reducing surface noise, paving the way for novel listening
practices and the constitution of new forms of sociabilities and subcultural
groups (such as the high-fidelity aficionados of the second half of the 1950s
and early 1960s[90]). The new plastic records provided an extended listening
time, but also had an extended physical durability and near-indestructability.
Vinyl records, which cannot be recycled and do not organically degrade,
geologically persist as 'toxic residue', becoming part of a 'fossil layer of
disused gadgets and electronics'.[91] Organic matter, such as clay and shellac,
'has its moment. It has history'.[92] Nearly indestructible synthetic plastics, on
the other hand, 'will never be over'.[93] They anticipate a different, extended
form of temporality – one which (measured against the brevity of human
existence) appears to be eternal.

The aura of vinyl in contemporary culture could be conversely described
as a radioactive effect. For media theorist Yves Citton, the mediatic past
exists 'radioactively' in that it continues to influence, in often implicit yet
transformative manners, the present: mediatic contents, in the moment of
their re-circulation, inexorably contaminate the contemporary.[94] Citton
links radioactivity and retroactivity so that the two are eventually indis-
sociable from one another: mediatic objects offer an endless 'feedback to
the future'. Deleuze's formula of the 'radioactive fossil'[95] may be employed
to describe how alluringly obsolete media technologies physically energise
– or irradiate – the present. Here, Deleuze's fossil seems to converse with
Benjamin's fetish in that both 'have in common a disturbing light, an eerily
beckoning luminosity'– inviting 'the viewer to excavate the past, even at his
or her peril'.[96] Here, we may replace the term 'viewer' with 'crate digger',
for whom the dangers of phono-archaeology might be actual. Indeed, the
industrial transition from shellac to vinyl records, more than a material

89 Flusfeder (2015), 43.
90 See Keightley (1996), 172.
91 Parikka (2015), 141.
92 Leslie (2020), 16.
93 Ibid.
94 Citton (2017), 216.
95 Quoted in Marks (2000), 80.
96 Marks (2000), 81.

substitution and a cultural 'revolution', must also be resituated within a larger environmental shift – denoting the increased ubiquity of toxic plastics in everyday consumption practices.

More radically, in his 1989 essay entitled 'Biodegradables' (1989), Derrida, discussing the perseverance of nuclear waste, highlighted its '[e]nigmatic kinship' with 'the "masterpiece"',[97] inviting us to consider the cultural matrix in terms of toxicity, persistence and indestructability. The concept of haunting and hauntology, which he formulated a few years later, can be seen as partly stemming from his early theoretical engagement with residual materialities. There is a spectral quality to waste as it expresses the continuous activity of the past leaking into the present. Derrida's framing of waste offers a reflection on historical contamination, latency, responsibility, and inheritance. Indeed, no detritus, no matter how deeply buried, may be deactivated – this applies to material as well as ideological leftovers. Within the Derridean framework, no complete elimination or effacement through recycling may occur. Rather, everything which once was persists through instances of mutation and deformation, implicitly energising the contemporary.

In his autobiographical reportage penned in 2015, novelist David Flusfeder has further reflected on the terminological and psychological disjunction between the high cultural valuation of vinyl on the one hand, and its often-unacknowledged hazardous materiality on the other. With their high lead content, the PVC pellets massively imported from Thailand and China to manufacture records have been classified as toxic for at least fifty years and represent 'the most environmentally pernicious plastic in use'.[98] In a semantic break, the term 'vinyl' has become separated from PVC, acquiring a mythological quality and autonomy of its own (in the Barthesian sense).[99]

Recyclability vs. plasticity

In his ecomedia study of recording formats (which he calls 'eco-sonic media'), Jacob Smith depicted the shellac era as that of 'the Green Disc network'. Smith

---

97 Derrida (1989), 845.

98 Maxwell and Miller (2012), 59; see also Flusfeder (2015), 45–46; Smith (2015), 35. As well as containing lead, PVC is derived from a toxic production process relying on high amount of chlorine.

99 In the first half of the 1970s, the British trade magazine *Music Week* published a series of articles dedicated to the poisonous impact of vinyl on health, mapping out, in a prescient manner, its carcinogenic effects (for both factory workers and consumers of recording sound).

draws a neat distinction between the non-polluting, recyclable gramophone and the toxic PVC LP which came to replace it, intimating that a return to bioplastic discs might constitute a step towards a more sustainable recording industry.[100] The Green Disc era, he claims, was 'a relatively low-impact and sustainable infrastructure' relying on a convivial process of manufacturing in which nature and human collaborated.[101] It is tempting to celebrate the apparently eco-friendly, nontoxic and renewable character of shellac. Yet, while shellac could theoretically be processed in a 'green' manner with no or little machinery, shellac factories were hazardous environments. Additionally, many toxic substances could be – and often were – involved in the pressing of 78rpm records (including arsenic and cyanide of potassium), and processes of both shellac and gramophone manufacturing were damaging rather than 'convivial' (see Chapter 2).

While Smith's interpretation provides a number of thought-provoking insights into the planetary dynamics of the early recording industry, the Green Disc narrative leaves many of its aspects unaddressed, notably in terms of power relations. The colonial undercurrents and power dynamics implicit in the early transnational record industry are referred to it an oblique and peripheral way, and the agency of the 'humble lac insect' appears to be overemphasised.[102] The flat network conceals the colonial logic which latently informed the early phonographic industry, and which the two World Wars acutely brought to the surface again. An ecomaterial account of the recording industry must therefore acknowledge the long and unsettled history of colonial contact, exploitation, and conflict lying captured within the grooves of the gramophone record. It may be argued that the early disc continues to '[raise] questions about trade relations between the Global North and the Global South': a return to shellac would arguably 'require a sensitivity to postcolonialism and neocolonialism that seems beyond the conscience of the capitalist recording industry'.[103]

It may be no mere coincidence that the first vinyl records became commercially available exactly one year after the Partition of India, while in South Asia discs containing shellac acquired a greater longevity. Some sources suggest that discs containing shellac were pressed well into the

100 Smith (2015), 38.
101 Ibid., 39. It is difficult to second the claim that the shellac trade left 'a minimal carbon footprint': the transportation of the material by rail and ship from British India to Europe and North America points to another reality.
102 Smith (2015), 39.
103 Devine (2019a), 80.

1990s.[104] The Dum Dum pressing plant began manufacturing vinyl records in the late 1950s – issuing its first 7-inch EPs in late 1958 and its first LP records the following year –, although these were initially aimed at the export market.[105] There is no clear rupture between shellac discs and vinyl discs: they belong together as the two sides a single history. Although they superficially exist in a relation of linear continuity (with one format 'replacing' another), it may be more productive to describe their relation as one of interference, superimposition or mutual parasiting.

Additionally, it is worth remarking that the act of recycling is not seamless or without a cost: 'recovering' an obsolete media material cannot occur without acknowledging, at the same time, how it genealogically came to be. To recycle may be to liberate ghosts or to allow for them to be, while recognising that phantoms will never be made voice and flesh again. If media cultures are haunted, I would like to suggest they can also be approached as plastic, which yields very different implications. If we return to Malabou's conceptualisation of destructive plasticity, plasticity also – and perhaps first and foremost – signals a lesion or a radical break (rather than a temporary stuttering) in the chain of becoming from which no return is possible. In the realm of objects, a material recovery may be possible, yet physical recycling cannot coincide with a recovery of times and lives past. On the one hand, materials exist in history, produce history and to some extent are produced by it. On the other hand, they also yield a reversible, elastic or cyclical potential which eventually makes them irreducible to the logic of historical time (and resistant to recuperation). In other words, materials which 'assimilated' history (becoming both the substrate and the medium of historical negotiations) cannot be assimilated to it or are not sufficient to redeem it. In some ways, as most systematically explored in the arts (which constitute the first field of plasticity), the recuperation, recirculation and transformation of obsolete media never mark a return of the past – not even as an approximation. There may be no getting nearer to the grain of what has been. And sometimes even materials resist recycling and remain mutely, irreversibly wasted. If we follow Malabou, not everything can be redeemed. In 1990, returning from a trip to the Golan Heights, Scottish poet Iain Crichton Smith wrote about the disaster still to come: 'Something is waiting for us in history like an unexploded mine, / like the crux of a scholarly book'.[106]

104 Berenbaum (1995), 122.

105 Kumar De (1990), 51. Amongst the first LP records to be manufactured in India were those of Ravi Shankar and Ali Akbar Khan.

106 Crichton Smith (1990), 119.

While the poet awaited the future past, it may be argued that Malabou
writes about the moment that comes after waiting: the mine has exploded
and meaning, which cannot be repaired or retrieved, must be created again
from scratch – without the image or the guidance of the 'past'. New media
theorist Alexander R. Galloway, examining the implications of Malabou's
plasticity for processes of reading, insists that a 'plastic reading' evidences
not the pre-existing structure (or foundations) of a text but bears witness to
its transformation (and liquidation) in real time. Structure therefore emerges
as 'a trace or fossil left behind by the text as it changes'.[107] Plastic reading
'comes *after* deconstruction' and is 'regenerative':[108] it doesn't retrospective re-compose, excavate or re-cycle a latent structure (in the way that a
reconstructive archaeology might do) but recognise the impossibility of a
return. The 'same' cannot take place again: what paradoxically remains is
a series of alterations.

Malabou's concern with radical transformation is helpful to deemphasise
and critically re-examine ecomedia theory's messianic trope (after Benjamin
and Derrida) of the spectral and the deferred – as well as its idealising 'hope
in the past'. The ghost has acquired a strong romantic potency in media
studies, yet a shift of emphasis from 'recyclability' (return, reversibility) to
'plasticity' (irreversibility) may invite us to reflect on moments of unbridgeable difference and consider that which, perhaps, cannot be repaired or
salvaged. On a superficial level, hauntology posits the existence of a contact
zone between materiality and history where, on a micro-material level, dust
'makes our historical ground and can be forensically examined' to reveal
'an archive of contamination, of violence, of repeated inequities that occur
on the surface'.[109]

The concept of plasticity, however, recognises that the temporality of
materials radically differs from the temporality of history – and though the
two are intimately entwined the former cannot be recuperated to retrospectively 'repair' the latter. It is always already too late. The retrieval and
recycling of obsolete (media) materials cannot offer us a return or a promise,
if only because the (ecological, human, etc.) damage has been irreparably
done, once and for all. It may be that which is chemically transformed, as
Esther Leslie underlines, is historically 'irretrievable':[110] metamorphoses
signal moments of irreversibility, loss and new departures.

---

107 Galloway (2012), 11.
108 Ibid.
109 Leslie (2020), 18.
110 Ibid., 4.

Bibliography

Affelder, Paul. 1947. *How to Build a Record Library: A Guide to Planned Collecting of Recorded Music*. New York: E. P. Dutton & Co. Inc.

Anon. 1942a. 'Shellac Order Strikes Disc Production'. *Broadcasting* 22 (16): 10.

Anon. 1942b. 'Transcribing Firms Discount Effect of WPB Shellac Order'. *Broadcasting* 22 (16): 10; 53.

Anon. 1956. *Shellac*. Angelo Brothers Limited: Calcutta.

Ballin, Anita. 1996. 'Women's Work in the First World War'. In *Women in Industry and Technology from Prehistory to the Present Day*, eds. Amanda Devonshire and Barbara Wood, 235–241. London: Museum of London.

Bauman, Zygmunt. 2000. *Liquid Modernity*. Cambridge: Polity Press.

Berenbaum, May. 1995. *Bugs in the System: Insects and Their Impact on Human Affairs*. Cambridge, Massachusetts: Perseus Books Blake.

Blake, Eric C. 2004. *Wars, Dictators and the Gramophone, 1898–1945*. York: William Sessions Limited.

Brittain, Vera. 1986 (1933). *Testament of Youth: An Autobiographical Study of the Years 1900–1925*. London: Penguin.

Broven, John. 2009. *Record Makers and Breakers: Voices of the Independent Rock'n'Roll Pioneers*. Urbana and Chicago: University of Illinois Press.

Brown, George I. 1999. *The Big Bang: A History of Explosives*. Phoenix Mill, Thrupp, Stroud: Sutton Publishing.

Citton, Yves. *Médiarchie*. 2017. Paris: Editions du Seuil.

Crichton Smith, Iain. 1990. 'On the Golan Heights'. In *High on the Wall: A Morden Tower Anthology*, ed. Gordon Brown, 119. Newcastle upon Tyne: Morden Tower / Bloodaxe.

David, Allison Matthews. 2015. *Fashion Victims: The Dangers of Dress Past and Present*. London: Bloomsbury.

Derrida, Jacques. 1989. 'Biodegradables: Seven Diary Fragments'. Translated by Peggy Kamuf. *Critical Inquiry* 15 (4): 812–873.

Devine, Kyle. 2019a. *Decomposed: The Political Ecology of Music*. Cambridge, Massachusetts and London, England: The MIT Press.

Dobell, Eva. 2006. 'In a Soldiers' Hospital II: Gramophone Tunes, 1919'. In *The Penguin Book of First World War Poetry*, ed. George Walter, 208. London: Penguin.

Ernst, Wolfgang, 2013. *Digital Memory and the Archive*. Minneapolis and London: University of Minnesota Press.

Fauser, Anegret. 2013. *Sounds of War: Music in the United States during World War II*. New York: Oxford University Press.

Flusfeder, David. 2015. 'Vinyl Road Trip'. *Granta* 138: 29–57.

Gaisberg, Fred W. 1948 (1947). *Music on Record*. London: Robert Hale Limited.

Galloway, Alexander R. 2012. 'Plastic Reading'. *NOVEL: A Forum on Fiction* 45 (1): 10–12.

Harris, Mark. 2017. 'Turntable Materialities'. *Seismograf* [online]. https://seismograf. org/fokus/sound-art-matters/turntable-materialities.

Jones, Geoffrey. 1985. 'The Gramophone Company: An Anglo-American Multinational, 1898–1931'. *The Business History Review* 59 (1): 76–100.

Katz, Sylvia. 1985. *Classic Plastics: From Bakelite ... to High-Tech*. London: Thames and Hudson.

Klee, Paul. 1965. *The Diaries of Paul Klee, 1898–1918*. London: Peter Owen.

Kumar De, Santosh. 1990. *Gramophone in India: A Brief History*. Calcutta: Uttisthata Press.

Lange, Britta. 2015. '*Posterestante*, and Messages in Bottles: Sound Recordings of Indian Prisoners in the First World War'. *Social Dynamics* 41 (1): 84–100.

Leslie, Esther. 2020. 'Devices and the Designs on Us: Of Dust and Gadgets'. *West 86th: A Journal of Decorative Arts, Design History, and Material Culture* 27 (1): 3–21.

Lowe, Jacques, Russell Miller and Roger Boar. 1982. *The Incredible Music Machine*. London: Quartet / Visual Arts.

Malabou, Catherine. 2012. *Ontology of the Accident: An Essay on Destructive Plasticity*. Translated by Carolyn Shread. Cambridge, Malden, Massachusetts: Polity Press.

Marks, Laura U. 2000. *The Skin of the Film: Intercultural Cinema, Embodiment and the Senses*. Durham and London: Duke University Press.

Martland, Peter. 1997. *Since Records Began: EMI, the First 100 Years*. London: B. T. Batsford Ltd.

Mawani, Renisa. 2015. 'Insects, War, Plastic Life'. In *Plastic Materialities: Politics, Legality, and Metamorphosis in the Work of Catherine Malabou*, eds. Brenna Bhandar and Jonathan Goldberg-Hiller, 159–196. Durham and London: Duke University Press.

Maxwell, Richard and Toby Miller. 2012. *Greening the Media*. Oxford: Oxford University Press.

Melillo, Edward D. 2020. *The Butterfly Effect: Insects and the Making of the Modern World*. New York: Alfred A. Knopf. Ebook.

Mosse, George. 1991 (1990). *Fallen Soldiers: Reshaping the Memory of the World Wars*. New York: Oxford University Press.

Osborne, Richard. 2012. *Vinyl: A History of the Analogue Record*. Farnham, Burlington: Ashgate.

Parikka, Jussi. 2015. *A Geology of Media*. Minneapolis and London: University of Minnesota Press.

Parry, Ernest J. 1935. *Shellac*. London: Sir Isaac Pitman & Sons, Ltd.

Pollard, Anthony. 1998. *Gramophone: The First 75 Years*. London: Gramophone Publications Limited.

Radano, Ronald and Tejumola Olaniyan, eds. 2016. *Audible Empire: Music, Global Politics, Critiques*. Durham and London: Duke University Press.

Read, Oliver and Walter L. Welch. 1976. *From Tinfoil to Stereo: Evolution of the Phonograph. Second Edition*. Indiana: Howard W. Sams & Co.

Roy, Elodie A. 2018. 'Worn Grooves: Affective Connectivity, Mobility and Recorded Sound in the First World War'. *Media History* 24 (1): 26–45.

Rust, Brian. 1975. *Gramophone Records of the First World War: An HMV Catalogue 1914–18*. Newton Abbot: David & Charles.

Scheer, Monique. 2010. 'Captive Voices: Phonographic Recordings in the German and Austrian Prisoner-of-War Camps of World War I'. In *Doing Anthropology in Wartime and War Zones: World War I and the Cultural Sciences in Europe*, eds. Johler Reinhard, Christian Marchetti and Monique Scheer, 279–309. Bielefeld: transcript Verlag.

Smith, Jacob. 2015. *Eco-Sonic Media*. Oakland, California: University of California Press.

5. Shards: Waste, obsolescence, and contemporary remediations

Abstract:

Chapter 5 attends to the recurrence and persistence of shellac in contemporary art and design, describing its visual and material remediation by contemporary practitioners based in Germany, France, Britain and the US. Their works are notably discussed in relation to the *Broken Music* exhibition (1988–1989), first shown in Berlin, which marked a critical turning point in what could be called 'gramophone art'. Throughout, the chapter discusses the importance of embodied modes of knowing for the exploration of materiality – and revives François Dagognet's invigorating plea for an ontology of (neglected) materials.

Keywords: contemporary art, design, phonography, waste, recycling, shellac

This last chapter retraces some of the afterlives of shellac – and gramophone records – in the contemporary era, at a time when it has fallen into obsolescence as a media material but continues to be an important symbolic resource or matrix in sound and visual art.[1] It discusses the works of contemporary artists (or, to use a helpful German coinage, 'research artists') and designers who have explicitly and systematically engaged with shellac through the knowledge of its deep media-material history – while moving beyond the rigid confines of historical retelling. In doing so, the chapter also highlights what multidimensional and intersensory modes of thinking may contribute to our understanding of media – in other words, it acts as a supplement to what came before, allowing us to 'actualise' the history of the material. This is not to say, as already noted in Chapter 4 in relation to the

---

1   It still has currency, however, in a number of other industrial contexts and is being used, for instance, as a vegan fruit wax and a gel nail polish.

Roy, E.A., *Shellac in Visual and Sonic Culture: Unsettled Matter*. Amsterdam: Amsterdam University Press, 2023
DOI 10.5117/9789463729543_CH05

theses of Catherine Malabou and Esther Leslie, that a material excavation can ever be equated with a strict 'return' to previous historical stratas.[2] Despite the effective endurance of materials, no repetition is ever possible: the previous chapters have insisted on the plasticity and transformational potentials of shellac to illuminate, in turn, the mutability and fragility of cultural formations themselves. In what follows, I would like to argue that contemporary sound and visual artists (not unlike their interwar predecessors discussed in Chapter 3) fashion valuable ways of hearing, thinking about and looking at the surfaces and substrates of the contemporary era. Their close, intimate involvement with the materiality of objects and archives may allow them to coin in turn a more expansive dialogue with the world than conventional academic researchers may feel authorised to form. This work is without end.

In the examples analysed below, the processual nature of artistic experimentation is highlighted. In such a conception, the discrete artwork appears not as the final culmination or completion of a practice – not as an autonomous, closed-off *objet d'art* – but rather as a temporary, emergent shape in a long, open-ended chain of experiments and material operations. I would like to recover here François Dagognet's pioneering (but sadly overlooked) contribution to the field of ecomaterial thinking. For Dagognet, a student of Gaston Bachelard, experiential artistic practices constitute another way of knowing – a knowledge from the inside, which often escapes verbalisation and the flattening logic of generalisation and objectification. His *Ecological Philosophy*, published in 1997, was one of the first philosophical studies dedicated to waste and excessive materialities. The author argued for an 'ontologisation of materials',[3] and proposed to integrate discarded materials and processes of making into philosophical thinking.

The ontologising project first outlined by Dagognet in the 1990s was most systematically developed by Jane Bennett in her book *Vibrant Matter: A Political Ecology of Things* (2010) where she reflected, amongst other things, on the affective possibilities of waste. For Bennett, litter constitutes a form of 'vibrant matter' which 'exhibit[s] its thing-power' and provocatively 'issue[s] a call': yet this doesn't mean that the call can be fully heard or answered. Rather, as Bennett notes, it is indicative of 'something more than it is possible to know'.[4] It follows that materiality cannot be limited

---

2   In archaeology, digging is inevitably disruptive as it is predicated on the destruction of previous layers. One needs to destroy what is here to understand what was there.

3   Dagognet (1997), 161; my translation.

4   Bennett (2010), 38.

to the realm of representation or to the representable. This has important methodological implications. What is therefore implicitly and continually at stake is the fashioning of decentred and nomadic, non-authoritative modes of philosophical writing – in the middle of things. This in-betweenness doesn't mean that no moral positioning, no foundation or no perspective may be adopted – or that a total withdrawal of and from conceptual forms is envisioned. It simply suggests that the dynamic principle must be restored and reinscribed within the philosophical enterprise itself.

Dagognet's ontology of mutable materials – he called it a 'materiology' – took the specific form of a (militant) revaluation of the 'insignificant' and the 'banned',[5] leading him to recuperate mesomorphic substances such as grease, tar and fat (which kept reappearing for instance in the work of Joseph Beuys, one of Dagognet's case studies).[6] Shellac belongs to this category of marginalised substances. In the realm of art, it is not valued or enhanced as a 'noble material', but often constitutes the mundane, hidden substrate of a painting – or its superficial finish (or varnish). US painter, sculptor and printmaker Jim Dine (1935–) describes it as 'traditionally a non-art material like tar or cement', explaining how he 'like[s] using things that are used by "workmen" for handmade art'.[7] Shellac recurs through the works he produced in the 1970s, 1980s and 1990s. Dine – nourishing a life-long fascination with tools as well as with ordinary, non-art materials – recalls first encountering it at home as a child, observing his father 'shellacking unfinished cheap pine furniture'.[8] From the 1970s onwards, he produced several mixed-media collages, paintings and prints which took the material as their points of material and symbolic departure. These include *Shellac Orientale* (1973–1974), *Shellac on a Hand* (1986), *The Side View in Florida* (1986) and *Shellac and Candy* (1993), whose titles obliquely recalled the prevalence of shellac as a confectioner's glaze, its geographical origins, or yet again its early ornamental uses.[9] Other painters have used shellac as a way of creating thickly layered, sculptured surfaces. Anselm Kiefer (1945–) often worked with the material (and with other organic substances), his monumental paintings producing a tangible sense of weight, obstruction and trapped

5   Dagognet (1997), 14; my translation.
6   A mesomorphic material is defined by its indeterminacy; it exists halfway between the solid and liquid states.
7   Personal communication, March 2022.
8   Ibid. It is worth recalling here that Dine's father ran a DIY shop.
9   Although there exist some resonances between them, it must be noted that Dine's 'shellac' works – discretely punctuating his long career – were never intended as a series as such.

energy – suggestive of a disintegration.[10] Decomposing or wasting materialities – including fragments and shards – may be 'one of the richest
areas of materiality'[11] because they present themselves as time in action
(or unstoppable becoming).

Dagognet's 'materiologists' (to reuse the term he coined to describe phi-
losophers undertaking research into materials) are close to Régis Debray's
mediologists: they posit no hard rupture between meaning (values) and
media (vectors). Dagognet's materiology may thus offer us the missing
link between material culture and media studies. It posits the residue as
that which resists and cannot be fully or successfully integrated within a
system – be it aesthetical or cognitive. As such, waste resists absorption and
recuperation. The question of waste, as understood by Dagognet, ultimately
draws our attention to the precariousness and artificiality of theoretical
constructs and to the necessity of in-between positions. Such positions
are uncomfortable and fragile, yet they might keep the work of thinking
alive, and forever emergent.[12] It follows that interdisciplinary thinking
cannot be reduced to an autonomous academic exercise. Rather, it may
constitute itself as an everyday practice – a striving toward that which always
already resists disciplining and cannot reach a term. In the end, waste, in
its most capacious sense, may be conceived less as a problem to be solved,
eliminated, repressed or resorbed, than as a positive, differential site of
living, feeling and thinking. It gives itself as an ambiguous, undisciplined
and uncontainable ground. This ground, however, may allow for movement,
failure, and experimentation to emerge again – in provisional yet vibrantly
necessary forms.

While waste has often been pushed at the margins of philosophical
enquiry, a reflection with and about wasted materialities lies at the core of
contemporary aesthetic practice. Much of twentieth century western art,
starting with Surrealism and intensifying in the wake of the Second World
War, has been concerned with the realm of the shapeless or the formless,
or what art theorist Nicolas Bourriaud terms the 'exform'.[13] It would be
impossible to give an exhaustive account of all the visual artists – including
painters and filmmakers – who have worked with the theme and the medium
of trash. The German painter Joseph Beuys, for instance, famously redeemed

10   Shellac was also a key material for British artist John Virtue (1947–), contributing to the
wistful severity of his monochromatic landscapes of post-war London.
11   Dagognet (1997), 26; my translation.
12   In keeping with Lisa Baraitser's thoughts on transdisciplinary hybrids. See Baraitser (2017).
13   See Bourriaud (2017).

sticky, organic materials such as fat, beeswax, or honey and reclaimed household objects. In a more symbolic way, underground filmmaker Jack Smith extensively reassembled unwanted stock footage, while Andy Warhol redeemed the cheapness of commodity culture, lending auras to discarded pop idols. With his memory boxes, Joseph Cornell transformed found detritus into delicate monuments. The immense list would also include the works of Robert Rauschenberg, Nancy Rubin, Gabriel Orozco, Ed and Annie Kienholz, Chris Casady, to name but a few of them. Each of these artists, in widely distinctive and different ways, operated a visual recuperation and revaluation of both consumer leftovers and unwanted matter. At the same time, by redeeming the repressed and forgotten materials of culture, they concretely enacted a genealogical or historiographical project which could be read alongside with Walter Benjamin's recuperative messianism.

The domain of art constitutes a particular source of inspiration in the ecological philosophy outlined by Dagognet. Art is seen as 'preceding' – without replacing it – the moment of philosophical elaboration. Dagognet doesn't suggest that art may – or should – articulate, formulate or solve philosophical problems. Rather, art may precede, energise or activate philosophical discussions – while operating in a radically different realm. Here, precedence is understood less as a temporal category than as a substrate or matrix. As well as redeeming leftovers, the work of Dagognet therefore attentively interrogates experiential form of knowledge and their undertheorised relation to philosophical conceptualisation – showing how, in the design process, the body may slowly become a mediating object or epistemic instrument in itself: simultaneously effacing and affirming itself, alternately acting as a conduit and as an obstacle. Dagognet insists that embodied (or poetic) forms of investigation produce 'forms of tacit knowledge': the latter may not oppose 'objective' or scientific knowledge as much as they pre-exist (or anticipate) it. They shouldn't be artificially devalued or repressed.[14] Accordingly, media materialities may be accessed through poetic knowledge (from the Greek *poieisis*: to make).

In what follows, shellac is envisioned as a form of resistant cultural surplus – a material and affective remainder which may (or may not) be transformed. Processes of wasting and situated practices of 'recycling' both artefacts and stories are highlighted. The act of recycling cannot be understood as a simple recirculation. This chapter is concerned with the works of artists who, at the time of writing, are actively engaging with the history of shellac and early phonography as a continual 'happening'.

14   See Adamson and Harris (2017), 18.

Although historical reflection – and, in nearly all cases, energetic archival research – inform their practices in important ways, they seek less to repeat or retell an historical episode than to express it today *as if for the first time*. As such, what they offer is less a totalising narrative than a material story of difference *and* persistence (where the relationship between the two terms is ceaselessly renegotiated).

A useful distinction can be drawn between narratives and stories, where the former is 'about continuity within experience' while the latter 'entails a process of differentiation within continuity': this means that '[s]tories entail narratives but also their contradictions, renegotiations, and their renewal'.[15] Stories – which acknowledge and encompass heterogeneous elements – thus have the capacity to alter or decentre the narrative whole (see Introduction). Accordingly, this chapter addresses a selection of discrete works created in the early twenty-first century – each of them responding in intimate, urgent and idiosyncratic ways to past shellac cultures, therefore proposing novel insights and individual ways of acknowledging, inhabiting and continuing the 'living story' of shellac. This necessarily partial selection, focusing on work produced in Germany, France, Britain and the US, notably discusses the practices of sound and visual artists including Christopher Dorsett, Dinah Bird, and Graham Dunning, and of the design duo Studio Lapatsch|Unger.

The chapter notably draws from individual conversations and interviews carried out in person and remotely in the 2017–2021 period. Although the selection inevitably contains a subjective element, it was made with a view of illuminating a constellation of approaches – driven by different, and sometimes divergent, sets of questions, interests and concerns (ranging from the biographical and the political to the ecological). The discussion of artist's works is organised according to three distinct categories: 'Phonographic heirlooms' (thematically focusing on of autobiographical memory, inheritance and erasure), 'Natural records' (focusing on the environmental and ecomaterial dimension of shellac), and 'Audible wastelands' (pursuing the environmental reflection on shellac by embracing themes of wasting and irreversibility). What brings these works together, however, is that all of them emphasise processes of making and storytelling rather than finished objects and closed-off narratives. There are also interpersonal links between my different case studies, with many of the artists and designers discussed in the following pages belonging to an informal community of shellac practitioners – a community which is physically organised around two

15   Jørgensen and Strand (2014), 60.

main sites, the German Phono-Museum (in St. Georgen im Schwarzwald) and the Paris Phono-Museum (France).[16]

This chapter begins with a broader panorama. Before highlighting individual art and design practices, it proposes a critical detour through the influential *Broken Music* exhibition (1988–1989) – the first exhibition dedicated to gramophone art – in order to reopen and actualise the important questions it posed, notably regarding the relationship between media objects, obsolescence, and 'unspent' or 'unexhausted' time.[17] This return to *Broken Music* is also important to situate the works of contemporary artists within the longer – and constellated – genealogy of what could be loosely termed 'gramophone art'.

Broken Music or the unexhausted record

The exhibition *Broken Music: artists' recordworks* was named after the fractured, recomposed records which Czech artist Milan Knížák's had begun producing in the mid-1960s.[18] Presented as part of the 1989 INVEN-TIONEN Festival for New Music, it opened at West Berlin's daadgalerie in 1988 before subsequently travelling to the Gemeentemuseum in The Hague (Netherlands) and Le Magasin in Grenoble (France), where it closed in February 1989 – a few months before the Berlin Wall came down, symbolically marking the progressive dismantling of the Eastern Bloc. The fact that the concept of the *Broken Music* show originated in the scarred and physically, as well as ideologically, split city of Berlin – where many of the artists discussed in Chapter 3 had met and exchanged ideas about new media before the Second World War – may not be a coincidence.[19] The

16   Having taken part in some of the events organised at the two Phono-Museums and personally met many of the artists I discuss, I situate myself in this text not just as an academic researcher but also as a member of this informal shellac community.

17   It must be noted that a second edition of *Broken Music* – entitled Broken Music Vol. II – opened in Berlin at the Hamburger Bahnhof in late 2022 as I was finalising this book. Time constraints haven't allowed me to include it in this chapter. I am grateful to Karin Weissenbrunner for sharing her recension of the new exhibition with me.

18   See Block and Glasmeier (2018 [1989]).

19   Berlin had also been the site of early gramophone-based happenings. In 1920, Dada composer Stefan Wolpe had launched a 'sonic assault' on Berlin, 'with eight Victrola gramophones playing Beethoven's Fifth Symphony at different speeds' (Harris [2017], n.p.). One decade later, at the 1930 Neue Musik Berlin festival, Paul Hindemith and Ernest Toch proposed a gramophone-based performance anticipating contemporary experimental practices: 'records were played simultaneously with live music, their pitch and timbre altered by speed variation' (Ibid.).

exhibition constituted the first major retrospective devoted to gramophone art, gathering the works of hundreds of international visual and sound artists from Marcel Duchamp and László Moholy-Nagy to Christian Marclay, Piotr Nathan and Joseph Beuys. Interestingly, the curators' interest lay with 'defective records' (William Carlos Williams's words) rather than with the dull promises of 'even, smooth reproduction'[20] offered by digital recording. In 1989, compact-discs became the prevalent listening format in the US and UK, where record sales had been sharply declining from 1982 onwards – a pattern of consumption soon to characterise listening cultures in the entirety of continental Europe.[21]

The celebration of defective records can simultaneously be read as a refusal to aestheticise consumer culture and its cult of pristine artefacts produced on the assembly line. As Lucy Lippard wrote in relation to 'New York Pop', 'Pop objects determinedly forgo the uniqueness acquired by time. They are not yet worn or left over'.[22] In other words, by denying the physical marks of ageing, pop art seeks to flatten historical time. It cancels depth and mediation by appearing as a perfect 'mirror' of the present (the celebratory gesture of the pop artist, however, may not be a completely uncritical or unironic one – and it may be argued that pop art is also a direct product of history and the historical imagination). There was nothing overtly pop about the *Broken Music* exhibition which (taking its cue from the pre-war writings of Adorno and celebrating early gramophone artists such as Moholy-Nagy) was unmistakably opening a dialogue with the pre-pop era, or with pop filtered through the ever-present memory of the Second World War. The curators (Ursula Block and Michael Glasmeier) were drawn to the disc as a uniquely wasted, useless or simply outmoded commodity – intuiting the moment when, divested of its commercial value, it could begin to truly speak (or perhaps stutter) for itself, articulating different messages about the contemporary. The pieces were original artworks in their own right and, simultaneously, anachronistic consumers' objects – latently retaining unspent or unexhausted 'time'.

Psychosocial theorist Lisa Baraitser, drawing from Michel Serres, proposes that 'a system can grow old without letting time escape, and through a process of temporal folding or kneading, unexpected contiguities and proximities can be made and remade'.[23] Although Serres, like Benjamin, was

20   Block (2018 [1989]), 9.
21   See Flusfeder (2020), 31.
22   Lippard in Stewart (2007 [1993]), 93.
23   Baraitser (2017), 44.

primarily concerned with 'the fate of "dead" ideas or texts'[24] and the possible surfacing of their latent meanings within new discursive assemblages, the image of 'folding or kneading' was directly inspired by the highly material gestures performed by artisans (in particular, Serres's reflection was guided by the image of the baker, indefatigably folding over the dough, 'trapping' different sequences of time within its folds). Artworks too may be understood as storing time as a 'dynamic volume' and therefore offering specific forms of 'counter-memory' where the stuff of history is transformed 'into a totally different form of time through the practice of vigilant repetitions'.[25] This is most explicitly the case with artworks deliberately reworking outmoded media artefacts and materials, at once remembering and re-inventing them. *Broken Music* demonstrated the many ways in which twentieth-century artists had plastically, symbolically and psychically reinvested – or recuperated – the ubiquitous yet obstinately opaque 'form of the phonograph record', a form which Adorno, writing in 1934 (under the humorous pseudonym of Hektor Rottweiler), had described with painstaking, quasi-archaeological precision:

> a black pane made of a composite mass which these days no longer name any more than automobile fuel is called benzine; fragile circular label in the middle that still looks most authentic when adorned with the prewar terrier hearkening to his master's voice; at the very center, a little hole that is at times so narrow that one has to redrill it wider so that the record can be laid upon the platter. It is covered with curves, a delicately scribbled, utterly illegible writing, which here and there forms more plastic figures for reasons that remain obscure to the layman upon listening; structured like a spiral, it ends somewhere in the vicinity of the title label, to which it is sometimes connected by a lead-out groove so that the needle can comfortably finish its trajectory. In terms of its "form," this is all that it will reveal.[26]

Adorno's forensic description (which resonates which Georges Perec's meticulous descriptions of household objects) deliberately defamiliarised the gramophone disc by presenting it as an inscrutable fetish – at once a familiar cultural commodity and an ancient carrier of indecipherable signs, to be interpreted anew. In this context, Adorno's essay could be read as an artefact

24   Ibid.
25   Ibid., 45.
26   Adorno (1990 [1934]), 56.

per se – an 'obsolete' text which was rediscovered and specially translated into English and French for the purposes of the exhibition, revealing new aspects of itself over fifty years after it was written.

As noted before, the principal focus of *Broken Music* didn't lie with records as desirable totems or frozen indexes of commodity culture (as they often were, for instance, in the flat, ironically stilted sensibility of pop art), but – in keeping with the programmatic title of the show – it pictured records as mobile and composite monads. Christian Marclay's now-celebrated floor of (vinyl) records – *Footsteps* – occupied one of the rooms. Many of the visitors, who had unreservedly embraced John Cage's *33 1/3* – a randomised composition for twelve turntables and one hundred records –, were hesitant to directly step upon (and therefore ruin) the smooth, exposed surfaces of the LPs. As co-curator Ursula Block observed, 'the conditioning to leave no traces on a record [was] too strong'.[27] But the exhibition provocatively asked how 'the creative use of records' – and their deliberate defacing and disassembling – may anticipate a wider 'break with conventional ideas, [...] a breakthrough into something new'.[28] Yet the breakthrough presupposes 'a kind of waiting', a withdrawal or suspension of both time and judgement: it constitutes 'a refusal of the new to be only orientated towards the future, and equally a refusal to simply "resurrect" the past',[29] echoing with Groys's conceptualisation of the new as originating from the cultural matrix of the old.[30] By drawing attention to ruined and reassembled media artefacts, the exhibition tentatively outlined possible narratives of repair, redemption and renewal, while questioning cultural habits – starting with listening itself.

In an autobiographical essay penned for the exhibition catalogue, René Block gives us poetical insights into the deeper substrate or pre-history of *Broken Music*. He recalls how, as a teenager in the mid-1950s, he was given a stack of shellac records – mostly film soundtracks – by a friend whose father owned a cinema.[31] The records – just as Marclay's records scattered on the floor in the exhibition – were sleeveless, and so 'scratched and dusty' that he would bathe them 'in warm soapy water and vigorously [rub] them dry'.[32] Unable to physically play them, Block washed the records 'over and over again', coining an idiosyncratic ritual to approach and understand them – so that, even silent, the caringly handled collection of records meant something.

27   Block (2018 [1989]), 9.
28   Ibid.
29   Baraitser (2017), 44.
30   See Groys (2014), 7.
31   See Block (2018 [1989]), 6.
32   Ibid.

Later, after he'd built himself a small, rudimentary record player, the fragile discs sometimes broke as he played them. The adolescent patiently glued the pieces back together to be able to keep listening. What mattered was less the physical intactness or integrity of the sound than the continuity of listening itself – understood as an everyday gesture of attachment, a material practice of maintenance mutually binding the sound object and the listening subject together – and which could also be read as an act of love in the widest, most impersonal and self-effacing understanding of the term.

In the aftermath of *Broken Music*, many further local and international cultural manifestations – often bearing less politically charged agendas – were curated to present the works of an increasing group of visual artists, performers and composers engaging with 78rpm discs and, more frequently, with long-playing records (as a shortcut for twentieth-century popular culture). Amongst these, we can cite the large retrospective *The Record: Contemporary Art and Vinyl* (2010) held at the Nasher Museum of Art at Duke University (US) which included audiovisual pieces by Laurie Anderson, William Cordova, Christian Marclay, and Ed Ruscha – to name only a few of them.[33] The same year, another large collective exhibition was launched at the Maison Rouge (Paris): it presented Guy Shraenen's vast collection of artists' records, focusing closely on the discs, inserts and record sleeves conceived by visual artists in the course of the 1950–1980 period.[34] In 2015, the collective exhibition *The Curves of the Needle,* named after Adorno's essay, took place in Newcastle upon Tyne (Baltic 39). Partially reclaiming the subversive sensibility informing the *Broken Record* exhibition, *The Curves of the Needle* – which featured the works of contemporary artists such as Jim Lambie (UK), Graham Dolphin (UK) and Bartholomäus Traubeck (Germany) – proposed to investigate records as 'a route to radical and conceptual practices'[35] rather than staging them as historical artefacts, collectors' fetishes or 'objects of desire'.[36] While the exhibition resisted the aestheticisation of commodities characteristic of pop art, strict boundaries between the record as a consumer good and as potential counter-cultural vessel appear to be difficult to establish.[37] One returns to Adorno's ambiguous fascination with the double life of the phonograph record, as it crosses over from commercial culture to avant-gardism, and back again. Indeed, the dates of the *Curves of the Needle* exhibition were selected

33 See Schoonmaker (2010).
34 Ratréma (2010), 201.
35 Pearson (2015), n.p.
36 Forty (1992 [1986]).
37 See Stewart (2007 [1993]), 92.

so that the exhibition could coincide with the annual celebration of Record Store Day (in April) – at the time, a record shop was located next door to the gallery, and the two spaces shared the same physical (and perhaps psychical) continuum. One recurrent feature of the exhibitions mentioned above, apart from the *Broken Music* one, was that they often failed to interrogate or renew the space of the gallery itself. Sound objects were there to be looked at, and though some of them were heard (more often than not in snippets) they weren't directly touched or interacted with.

Broken Record offered a concrete reflection on the nature of history as an unstable ground, introducing the surface of the record as a metaphor for the broken and damaged ground from which the present continuously and precariously emerged. It therefore urged listeners to acknowledge the very surface of history, its grain and its dust. Esther Leslie, in an essay on dust and devices beginning with a reflection of the fall of the Berlin Wall, reflects on the non-teleological, stuttering movement of history which 'moves [...] but does not seem to go forward'[38] – like a needle stuck in a groove. In *Broken Record*, the history of phonography was not presented as a catalogue of dead media but as a living memory where past machines mattered in and for the digital present. In doing so, *Broken Music* insisted on history as performance – made, maintained and potentially unmade in the sphere of the everyday. But history as performance is distinct from history as progress. The work of composition, decomposition and re-composition is never over – be it in music or in history. The remainder of this chapter discusses the works of artists and designers who have reengaged with the persisting materiality of shellac and gramophone records, both in biographical and impersonal ways.

Phonographic heirlooms

Christopher Dorsett: *Seven Chakras (Twice)*

In 2011, British curator, artist and academic Christopher Dorsett showed his series *Seven Chakras (Twice)* as part of the touring exhibition *Dust in the Mirror* inaugurated at Gallery North[39] in Newcastle upon Tyne (UK). *Seven Chakras (Twice)* was a series of graphite drawings – or 'over-drawings' – made on top of high-resolution photographs of home-recorded discs. The project was based on a collection of now-unplayable home-recordings Dorsett's

38   Leslie (2020), 4.
39   Gallery North is Northumbria University's staff-led gallery.

father had made in the years following the Second Word War at a time when he had set up, with his two brothers and an uncle, a 'voice letter' service called Replicords.[40] At the time, 'nearly everyone they knew had been separated from their relatives by six years of wartime upheaval':[41] sensing a commercial opportunity, Dorsett's father and his associates thought that 'postal records would make money at popular leisure locations where people gathered in large numbers'.[42] The service never took off but Dorsett's father made many test records. When Dorsett found the recordings, shortly after his father's death, he mistakenly took them for shellac discs, before realising (many years later) they were acetate records. I take Dorsett's initial mistake as an interesting slip, an idiosyncrasy which can be generalised: for any disc of the pre-vinyl era, independently from its actual composition, tends to be incorrectly labelled as 'shellac'. This cultural slip is not the only reason why I have chosen to include *Seven Chakras* (*Twice*) in the present study – a key reason is that the work, as described below, offers us a valuable prompt to theorise the entwinement of the sonic and the visual, and examine memory and erasure in relation to vulnerable media carriers.

Dorsett remembered first hearing the recordings of his father's voice as a child, when he had absorbedly played them in the family's attic 'on rainy days'.[43] By the time he became reacquainted with the collection of discs, however, they had become inaudible and unplayable – as irretrievable as the time when he'd first heard them. The discs were media that were 'not *anymore* media'.[44] What remained was a flat, tattered picture of sound – a record of what had ceased to be: the sonic signal could no longer be replayed (at least not in the way it was intended to be played). For Dorsett, the inaudible recordings became a fitting metaphor for the buried voice of his father as it lay contained in the disintegrating shell of the disc. The lost (or nearly lost) voice, however, anticipated a reflection on legibility, memory transmission and translation.

Rather than discarding the records as obsolete artefacts, Dorsett asked under which conditions the ruined discs may become mediated and operative again. How could they make sense? How could one recover (on an aesthetical or affective level) the mechanisms of playback and experience

---

40   Greek sound and visual artist Panos Charalambous has similarly worked with acetate home-recordings, notably in his exhibition *Aquis Submersus* (Athens, 2014). For a survey of Charalambous's practice, see Panopoulos (2018).

41   Dorsett (2020), 45.

42   Ibid.

43   Dorsett (2020), 45.

44   Parikka (2015), 37.

again the poignantly present moment of listening? It follows that Dorsett's investigation was not a strictly technical one, but was guided by an ontological questioning of what media are and what they may do. His aims were less to recover the objective contents and information but rather to restore (or create) the conditions for a sensory or phenomenological 'contact' with the artefact to take place. This notably explains why he did not follow the path of archival reconstruction, even though the sonic traces may have been digitally retrieved and replayed (according to a process similar to the one developed by the First Sounds team in California who were able to play back de Martinville's phonautograms in 2008).

In recent years, audio engineers have been developing ways of scanning damaged discs in order to visually 'repair' interrupted groove paths. Contemporary methods of sonic restoration extensively rely on digital imaging, alerting us to the centrality of optical methodology in the recovery of sonic cultures. The Smithsonian Institution, as Dorsett later learned, holds a machine which can 'recover sounds off damaged acetate records'. Yet, as noted above, physical restoration – and data reconstruction – were peripheral to Dorsett's enquiry. How could the mute records be made to function again – or be revitalised? The damaged home-recordings struck Dorsett as one-dimensional images of sound: the mute records, which could only be experienced visually, reminded him of the severe flatness of monochromatic photographs. This visual affinity led him to reproduce the records by photographical means. Dorsett's gesture unsuspectingly re-enacted Berliner's early forays – in the wake of Charles Cros's 1877 findings – into the photomechanical reproduction of sound and proposed that gramophone records could be engraved 'mechanically, *chemically*, or photo-chemically'.[45] Berliner illustrated further the idea of photo-engraving through a concrete – yet fictious – example, imagining how the photographical process may particularly benefit the Pope wishing to rapidly reach his fidels:

> Supposing that his Holiness, the Pope, should desire to send broadcast a pontifical blessing to his millions of believers, he may speak into the recorder, and the plate then, after the words are etched, is turned over to a plate-printer, who may, within a few hours, print thousands of phonautograms on translucent tracing paper. These printed phonautograms are then sent to the principal cities in the world, and upon arrival they are photo-engraved by simply using them as photograph positives.[46]

45   Berliner (1888), 437.
46   Ibid.

Where Berliner envisioned photography as a means of quickly mass-reproducing records in order to ensure the wide dissemination of a uniform message, Dorsett used high-resolution digital photographs (obtained through scanning the records) in order to further singularise the discs, magnifying their material imperfections and defects. In addition to the digital photographing of records (which created a distance from the original objects), another layer (and another distance as it were) was added by hand when the records were repeatedly drawn over in graphite, concealing and silencing the grooves. Dorsett physically retraced the unplayable grooves of the ruined records, using his pencil as a 'gramophone needle'. He recalls:

> Around and around my hand would go in mimicry of a needle cutting a groove in the acetate. These undemanding kinetics kept me drawing. I produced more and more versions, each whirling graphic 'thing' becoming a substitute for the defunct sound objects. It was a curious indexical endeavour that seemed to retreat from graphic expression. The results had to be exactly the same size as the records. They had to be just as flat as well as just as silent.[47]

The repetitive, free-hand graphite drawings infused the still images with a sense of mobility which recalls – without imitating it – the movement of playing records (Figure 5). The over-drawings – or drawings on photographs – were as unique as the home-recordings were. Over-drawing was inspired by the physically demanding domestic practice of varnishing furniture: it offered a means of, incidentally, re-lacquering or re-glossing the records (where glossing, etymologically, refers both to the physical act of shining, and to the practice of interpretation). It may be noted that Dorsett has often resorted to the practice of varnishing across his various projects, a technique which he learnt alongside many other art students as part of his formal art school education. But varnishing may be understood as a means of compensating for the scratched, damaged records: in this way, it appears as an exemplary gesture of care. Dorsett's practice may be compared to Gerhard Richter's overpainted photographs, to the thickly layered paintings of Frank Auerbach and Leon Kosoff, or even to the disappearing self-portraits of Roman Opalka, obtained from the subtraction (rather than addition) of layers.[48]

The over-drawings liberated (or remediated), in another form, the voice which had been silenced by the natural deterioration of the acetate discs.

47   Dorsett (2020), 46.
48   Roy (2016 [2015]), 18–19.

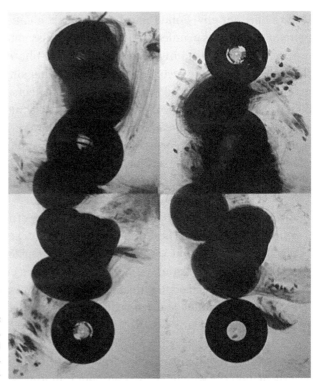

Figure 5: *Seven Chakras (Twice)*, Chris Dorsett, graphite over digital scans of recordings, 2011 © Chris Dorsett.

The iterative process of drawing could be compared to a form of recuperation where sound becomes poetically recovered as an image. It may be argued that the 'silence' of the drawings does more than paralleling the silence of the ruined records: it transforms mutability into an intelligible form of expression, where the drawing 'speaks' and becomes immediately coherent (the way the recording became coherent through the act of playback). What occurs is a transversal, trans-sensory operation where vision recovers – or rediscovers – sound. There is a process of displacement and relocation of meaning – a sensory reframing of the object. Here, the voice of the ruined records is redeemed in the sphere of drawing. Such a process can be seen as a form of trans-sensory poetic restoration and reparation which is very different from the detached type of technical restoration performed in the conservation departments of archives and museums. Finally, it must be noted that Dorsett insists that *Seven Chakras (Twice)* ultimately aimed at erasing the personal or autobiographical elements in order to find a common space of resonance. Yet what is visible in exhibiting the over-drawings is a direct visualisation of the repeated practice of playing records – with each instance of playing infinitesimally altering the groove as well as the shape of the story. They formed an interface between the initial home-recording and Dorsett's own version of it: in this sense, *Seven Chakras (Twice)* can be understood as a re-recording. The original sonic inscription, though completely buried, continues to noisily bleed through the drawings.

Dinah Bird: *A Box of 78s*

Another autobiography-driven project was the long-playing record *A Box of 78s* (2014) by British radiophonic artist Dinah Bird. In 2000, Bird inherited a leather box of shellac discs which had belonged to her recently deceased grandmother along with her great-grandfather's diaries. Bird's grandmother was born on the Gulf Islands (British Columbia) in 1910 and grew up on Salt Spring before the family settled in the United Kingdom in 1925. The records, which she collected as a teenager, travelled with her on the long journey from Canada to Europe. The collection, whilst presenting an unexceptional assemblage of 'classical music and opera hits of the day',[49] was remarkable in that the records originated from an array of international labels (including German, British and North American ones), illustrating the efficiency of record distribution networks at the time. Even a town as small as Ganges (BC) would have boasted at least one record shop, storing brands from around the world. Phonographic commodities can be described (by analogy with D. W. Winnicott's transitional objects) as 'transnational objects' which are 'created in cultural translation and transcultural movement'.[50]

 Bird decided to retrace the records' 1925 journey in reverse, taking them back to the places where her grandmother had first heard them as a young girl. Accordingly, '[u]sing [her] great-grandfather's diaries and daily notes on the weather as a guide', Bird 'played the records outdoors, on a portable gramophone in spots around Salt Spring where [her] grandmother and her family had picnics, played tennis or danced, and [she] recorded what happened'.[51] Whilst in Salt Spring Island, she also carried out interviews with her great-uncle, former neighbours and other inhabitants of the island. Later she interspersed the moments of playback with fragments of the interviews, producing a long-playing record out of the recombined layers of music, field recordings and stories. Although *A Box of 78s* was mass-duplicated,

---

49 Bird (2014), liner notes.
50 Marks (2000), 78.
51 Bird (2014), liner notes.

commercially issued and distributed, its meaning or surface could not be completely stabilised.[52] The dust never settled.

The new record, which resulted from the travelling of gramophone discs from Canada to Europe and back again, was not conceived as a static commodity but intended as a travelling radio programme. It contained a listening log encouraging participating radios on the relay run as well as individual listeners to complete the form (notably indicating when, where and by whom the record had been played as well as the frequency and '[w]eather on day of play'), and send it back to Bird. The involvement of radio stations replicated, expanded and objectified Bird's personal journey and intimate involvement with her grandmother's discs. Rather than a purely autobiographical project – and rather than a unified collection of records –, *A Box of 78s* becomes the site of a geographical and symbolic dispersion. One single copy *A Box of 78s* was shipped and relayed in experimental radio programmes across the world. Through the relay system, the record travelled from one radio station to the other, getting inevitably scratched and damaged in the process (echoing with projects such as Christian Marclay's 1985 work *Record Without a Cover*). Every superficial alteration and cut, however, contributed to adding a layer of symbolic depth to the recording so that, in Marclay's words, 'The vinyl record becomes a palimpsest that has a history of layered marks that you can't erase; incidental scratches become a natural part of the piece, not a mistake but integral to its meaning and composition'.[53] Random events are thus retrospectively endowed with an aesthetic legitimacy, becoming integral part of the composition.

A Box of 78s functions as a series of (re-)recordings within a unified record-ing, operating a series of remediations. The first one is a material remediation of the gramophone disc as a media format (from shellac to vinyl).[54] The second instance of remediation could be understood as a remediation (or re-forming) of reality itself – where her grandmother's actual record collection, alongside with her grandmother's father's diary, becomes the basis for a fictional retelling of bygone (and possibly romanticised) episodes of family life. Whilst the first side of the record contained a radio play or *horspiele* (combining original re-recordings of the shellac discs with ambient sounds and newly recorded interviews), the second side featured 'a series of locked grooves made from field recordings from the trip'. Sampling – and

---

52   The album was released on the German experimental label Gruenrekorder.
53   Estep and Marclay (2014), 39.
54   Visually, it may be noted that the sleeve of the new album – housed in plain brown cardboard – was reminiscent of the earlier gramophone sleeves.

sonic loops – are particularly appropriate to suggest a sense of cyclicality and closure. At the same time, no full 'return' or 'iteration' may take place: what the new record captures is a moment of displacement.

The project can further be understood as part of a larger process of mourning where the act of partial repetition may ultimately lead to both an understanding of the past and to an emancipation from it. As a new artefact, *A Box of 78s* 'buries' the original collection of shellac discs: the latter become symbolically suppressed and altered in combination with other recordings, becoming sonically unrecognisable. Yet, despite their immensely different aesthetics, there is no hard and strict rupture between the original records belonging to her grandmother and Bird's new composition as it supplements the initial collection, preserving it through reinvention. At the end of the project, all the listening logs gathered by Bird were collectively exhibited, while 'the remains of the records [were] played for a final time before being laid to rest'[55] – a formulation presciently pointing to the necromantic imagination of the phonograph, and recalling Christopher Dorsett's visual 'burying' of his father's records as he repetitively drew over their photographed surfaces. Both Bird and Dorsett, through their reinterpretations of family heirloom, chose the route of invention and deconstruction rather than that of faithful replication. Neither of them was interested in merely 'reissuing' or 'reproducing' obsolete artefacts, but in how these artefacts significantly conversed with the present moment. Instances of retelling can be understood as paradoxically means of preserving the past: they are the equivalent of a storyteller adding, altering or subtracting an item every time she retells a story.[56]

Natural records

Studio Lapatsch|Unger: *The Forgotten Collection*

While the two previous case studies emphasised the relationship between artefacts and autobiographical memory, Studio Lapatsch|Unger achieves a more impersonal type of memory work – recovering abandoned types of

---

55   Bird (2014), liner notes.
56   Bird continues to explore the mobile geography of gramophone records and their relationship to processes of displacement, identity and loss. In her follow-up 2022 project *Cent Mille* ('One Hundred Thousand') the artist travelled to India to visit shellac factories and interview shellac workers. I am revising my manuscript as Bird is returning from India, and regretfully am not able to include *Cent Mille* in this chapter.

materials and recipes for plastics devised in the eighteenth and nineteenth centuries – in the hope of drawing attention to a disappeared 'collective world of experience'. Studio Lapatsch|Unger, formed by designers in Anja Lapatsch and Annika Unger in Berlin in 2016, describe themselves as 'physical narrators' – another expression for material storytellers – who 'use design as a narrative agency of things'[57] and whose work is guided by the materials themselves rather than by preconceived notions of form. As such, they offer us a concrete example of what following the materials may signify in the context of a design practice.

With *The Forgotten Collection*, appropriately displayed at the Bauhaus Dessau Foundation in 2017,[58] they focused on largely discarded animal and plant-based polymers including birch pitch, rosin, dammar, copal and shellac,[59] reflecting on the obsolete early plastics of cultural production. Combined with bamboo charcoal, shellac became the focus of Studio Lapatsch|Unger's second edition of *The Forgotten Collection* (2016), a project investigating discarded natural polymers. By 'investigat[ing] what is long forgotten', they invite viewers to 'think deeply about the modern modalities of materiality, time, value, production and reproduction' in the hope of 'creat[ing] a new ontological imagination'. Such an ontological imagination can be resituated within the context of François Dagognet's plea for the 'ontologizing' of neglected materials and of Gaston Bachelard's 'material imagination' (where the philosopher set out to question how the four elements – water, fire, earth and air – interrelated with the human imagination). Their project further echoes with the research into new materials initiated at the Bauhaus in the 1920s (see Chapter 3). In the course of the *Forgotten Collection* experiments – which went through two editions – they produced a series of deeply sculptural objects (including the carafe shown below; see Figure 6). Because of the materials they were made of, all of these objects ended up being a different shade of black – their ancient, charred aspect produced a superficial sense of kinship across the works, although each object carried with it the peculiarly meaningful accidents of its surface. In Lapatsch and Unger's rejection of 'the impeccable surface[s] of mass-production' – which could include the surfaces of new records –, we find echoes of Block and Glasmeier's dismissal

---

57 Lapatsch and Unger, personal communication, October 2021.

58 As part of the *smart materials satellites. Material als Experiment* exhibition which ran from July to October 2017 and stemmed from a collaboration between the Weißensee Academy of Art Berlin and Bauhaus Dessau Foundation.

59 Incidentally, almost all of these substances were used in the manufacture of gramophone records.

Figure 6: Carafe (shellac, bamboo charcoal), *Forgotten Collection*, 2nd edition, 2016 © Studio Lapatsch|Unger.

of perfect reproduction. Here, perfection – from the Latin *perfectionem* (suggesting something finished or complete) – appears as the undesirable antithesis to life and movement. The asymmetric, almost crude-looking and unfinished objects emerging from Lapatsch and Unger's studio are new yet at the same time old and battered (partly because they are made of ancient materials according to long-discarded and imperfectly mastered techniques of production).

The irregular beauty of the objects produced by the duo could be understood according to the Japanese concept of *wabi-sabi* – which 'condenses an aesthetics of worn, dull, broken, and imperfect objects', relating to homemade and nonstandard objects 'that are somewhat ambiguous in their shapes'.[60] *Wabi-sabi* can be read as the counterpoint to 'today's sleek, mass-produced, industrial artifacts' with their 'perfect

60   Rabaté (2018), 99.

finish'.[61] Importantly, *wabi-sabi* cannot be reduced to a superficial, and therefore replicable, 'style' or design statement, but proceeds from a larger ontological discourse. It indicates a wider philosophical attitude to life itself, where differential processes of ageing, death, and decay are acknowledged – and even celebrated – in their everyday manifestation (rather than being hurriedly concealed, as tends to be the case in western culture and society, fetishising the new and supressing physical signs of ageing). When the term *sabi* first appeared in eighth-century Japanese literature, it exclusively denoted solitude, only acquiring the physical connotation of 'rusty' in the following centuries.[62] Literary scholar Jean-Michel Rabaté, discussing the poems and haikus of Bashō, writing in the second half of the seventeenth century, highlights their proto-Romantic sensibility where beauty 'derives from a sweet melancholia associated with the contemplation of objects that evoke impermanence'.[63] Here, the contemplation of worn artefacts is implicitly related to subjectivity, narratives of self-understanding and the melancholy realisation that everything must pass. Yet Bashō's writings should not and cannot be reduced to a westernised reading of them.[64]

Ruination, as well as the ontological reading of decay, were first conceptualised by the encyclopaedists in the eighteenth century (starting with d'Alembert's entry on 'debris, vestiges, ruins' for the *Encyclopédie* and continuing with Diderot's meditations on ruins in his Salon of 1787).[65] The Occidental cult of ruins – culminating in the Romantic period with the writings of Chateaubriand – differs from *wabi-sabi* in many ways (notably because it was infused with Christianity). Yet similarities exist: one of Chateaubriand's insights was to 'couple the ruins of nature to those of mankind',[66] evidencing their double belonging to both nature and culture. *Wabi-sabi* perhaps most closely resonates with art historian Alois Riegl's Romantically informed appreciation of the auratic 'age-value' of ageing artefacts and buildings at the turn of the twentieth century and their immediate affective appeal for '[a]ge value [...] addresses the emotions directly; it reveals itself to the view through the most superficial, sensory (visual)

61   Ibid., 101; Ibid., 98.
62   Ibid., 94.
63   Ibid., 96–97.
64   Importantly, they offer us a non-fatalistic conception of time as cyclical rather than irreversible and linear.
65   See Schnapp (2018).
66   Ibid., 160.

perception'.[67] Riegl's 'aesthetic of disintegration'[68] was to find a profoundly resonant continuation (or afterlife) in the writings of Walter Benjamin (specifically in his reflection on the displacement of the aura in the era of mechanical reproduction). What both the notions of *wabi-sabi* and age-value posit is the inextricable entwinement of matter and temporality, as well as an exploration of how they continually and mutually exceed (or compete with) one another. On the one hand, the past is 'crystallised' or petrified in the form of a three-dimensional object or container. On the other hand, it can also be seen as a surplus, continuously leaking out of its container – a process which can be accelerated when the object is purposefully crushed, dismantled, melted or reassembled (as was the case in the *Broken Music* exhibition).

The work of Studio Lapatsch|Unger differs from that of artists engaging with found or discarded objects of popular culture in that it is directly informed by the materials and their affordances, often starting with recipes and experimentations rather than with forms and models. The objects produced within the studio, however, can be easily mapped onto everyday artefacts. Some of them are even modelled after consumer goods (the first collection contains, for instance, a baseball cap). However, their unfinished appearance, their colour and density make them look and feel highly unfamiliar. Crucially, they may not be worn or consumed as commodities: they appear as unconsumable consumables. One is offered a glimpse into a parallel world of objects, suspended between the known and the imagined. Taken together, they form an imaginary counterpoint to the actual 'technofossils' of the Anthropocene era, defined as 'the sedimented debris-layers of our vast compressed cities [...] now constitut[ing] part of the geological and planetary record'.[69] Accordingly, Lapatsch and Unger's avowed desire to renew the 'ontological imagination' of forgotten materials – and notably to expose, rather than smooth, their archaically imperfect grain – can also be read as a critical revaluation of the material present. Lapatsch and Unger's approach to design is experimental, process-driven, speculative and – we may argue – political: they seek to rediscover the latent possibilities and alternative histories contained within materials itself – so as to tangibly invent (and, if it were, retrospectively salvage) a present which could have been and still could be, reclaiming the overlooked fringes of the modern project in a physical rather than metaphorical way.

67   Riegl (1996 [1903]), 74.
68   Iversen (1993), 34.
69   Thill (2015), 4.

Their exploratory approach deeply resonates with what Svetlana Boym once defined as 'the off modern', understood as 'a detour into the unexplored potentials of the modern project. It recovers unforeseen pasts and ventures into the side alleys of modern history at the margins of error of major philosophical, economic, and technological narratives of modernisation and progress'.[70] It is through experimentations with dysfunctional digital devices – rather than through strictly theoretical means – that Boym comprehended and erratically navigated the off modern. She playfully wrote about it as 'a performance-in-progress, a rehearsal of possible forms and common places. In this sense off modern is at once con-temporary and off-beat vis-à-vis the present moment. It explores interstices, disjunctures, and gaps in the present in order to co-create the future'.[71] The practice of Studio Lapatsch|Unger, incorporating discarded skills, recipes, and substances, expresses an unmistakeable 'discomfort with modernity' – a marked uneasiness with capitalistic modes of production – and a wish to participatorily engage with long-disappeared techniques and regimes of making (for the revival of past gestures may at the same time anticipate a slight yet decisive off-setting, an alteration and renewal of one's sensibility, a repositioning of one's body both in aesthetical and political terms).

Yet the return to pre-industrial practices, rather than signifying a nostalgic escape into the past, may simultaneously be understood as the difficult and uncertain carving out (or digging up) of a new present. In his cultural analysis of the new, Boris Groys insists that anachronism – which he defines, in simple terms, as 'prefer[ing] the old to the new'[72] – may produce a novel cultural paradigm because it critically re-examines the dominant rule 'requiring us to produce newness and, accordingly, form the radically new'.[73] Anachronism is an in-between – and therefore untenable – position which re-generates the past for the present: it follows that each 'renaissance' is equally a 'revival'.[74] Groys evidences the complex cultural mechanism by which every epoch adapts and filters the values of past eras, showing how the new is inseparably bound with tradition. He particularly reflects on the heterogeneous and amorphous realm of the 'profane' – defining it as that which lies outside the archival order and therefore could be understood as

70   Boym (2010), 1.
71   Ibid., 2.
72   Groys (2014), 7.
73   Ibid.
74   Ibid.

a form of waste.[75] Yet only the profane – a disorganised, excessive reservoir of discarded knowledge, materials, gestures, etc. – may serve as a potential origin of the new. Groys's recognition of the old as the source of the new echoes with Boym's off modern. Eventually, it may be suggested that *The Forgotten Collection* poses the practical question of how to acknowledge and inhabit the deep contemporary – across its heterogeneous and communicating chambers.

The emphasis placed by Lapatsch and Unger on material transformation (rather than on form and function) has implications for the conceptualisation of design as a discipline. The aims of the studio are not to produce finished or permanent objects – let alone to produce 'characteristic' or iconic artefacts. It could be argued that their practice, though deriving from an intimate engagement with materials, is simultaneously predicated upon a certain degree of depersonalisation or detachment, including a detachment from the notions of futurity and durability themselves. The object they create have no use-value and no future as commodities. Additionally, practices such as the one developed by Studio Lapatsch|Unger highlight the unfinished and processual nature of both design and matter, as being bound with temporality and processes of decay, effacement and potential renewal. In order to let the materials speak, it may be that the designer must remain silent and partially disappear – or strive to become a neutral or near-neutral medium. Accordingly, she can be best understood as an interpreter, revealing their concealed qualities, rather than as an originator or inventor of novel forms and objects as such. This resonates with Dagognet's conception of the artist as someone who humbly 'recognize, unfold or enhance a material'.[76]

Audible wastelands

The revaluation of obsolete, handmade manufacturing processes can be further observed in the works of contemporary sound artists and designers engaged in the artisanal making of records. The making of homemade records, relying on crude casting techniques, has almost became a cliché in the past decade and may be seen as an update on the phono-fetishism of the interwar period (see Chapter 3).[77] For instance, British sound artists

---

75 The profane realm contains 'everything that is valueless, nondescript, uninteresting, extra-cultural, irrelevant and [...] transitory' (Groys [2014], 64).

76 Dagognet (1997), 144; my translation.

77 For a survey of turntable materialities, see Harris (2017).

Michael Ridge and Ian Watson both revisited obsolete casting methods. While Ridge produced sound objects out of cast glue, Watson released a 7" inch record – *Only Surface Noise Is Real* (2014) – 'made from a silicon mould that was itself cast from a vinyl master'.[78] The title of the record alluded to the sedimented cultural valuation of discs as the most 'authentic' of all musical formats (implicitly devaluing digital listening as 'unreal'). In addition to the rediscovered artisanal casting of records, an opposite trend has been to employ ultra-contemporary techniques (such as laser-cutting) in order to carve grooves in a range of materials (as demonstrated in the works of design engineers such as Amanda Ghassaei and Kazuhiro Jo). In *Repetitive Movements* (2018), Sascha Brosamer manually made moulds of gramophone discs which he subsequently re-made out of oceanic plastic trash. These defective, homemade plastic discs are the exact opposite of the perfect (and perfectly impossible) shellac discs briefly fantasised by sound artist Robert Millis (see description below).

In *Repetitive Movements*, culturally valued recordings (originally pressed on shellac) get irreverently remediated as plastic rubbish. Rather than a virtuous, linear recycling, Brosamer offers a superimposition process where the recording can be heard simultaneously as indestructible and as destroyed. The recycled plastic discs suggest that repetition is best understood as a re-composition. Taking the term repetition literally, German-based sculptress and 'fabricator' Irene Pérez Hernández (based in St. Georgen im Schwarzwald) has used shellac as a source in one of her loop series (*Loop series IV: Endless Shellac*) to illustrate the seemingly infinite recirculation of the material. The loops, fashioned after the mathematical sign for infinity, metaphorically evokes the practice of 'listen[ing] to a record over and over again' while alluding to the broader recyclability of the material and to its inscription within a 'cycle of death and procreation'.[79] Her loops were based upon a rigorous understanding of the physical properties of the substance, and her material research began when she was commissioned to make shellac blanks (out of pre-existing, melted discs) by Berlin-based artist and educator Darsha Hewitt (as part of Hewitt's ongoing project *Protoplastic Prototypes* [2021–]). The blanks were subsequently sent to a record-cutting studio (Studio Alex Rex, Germany). The experience of making shellac discs – and the knowledge of the material's history – encouraged Pérez Hernández to reuse it in her own loop series (previously made of steel, glass and plexiglass, amongst other materials).

78 Harris (2017), n.p.
79 Pérez Hernández, personal communication, June 2022.

New York-based artist Jean Shin has also reflected on the ambiguous physical endurance of shellac in *Sound Wave* (2007), a monumental, petrified wave consisting of melted gramophone discs on a wooden armature. The wave was envisioned as somebody's record collection – recomposed and recoded as a wave of oblivion. While Pérez Hernández's loops draw attention to practices of circular listening, Shin's wave makes visible the obsolescence and vulnerability of musical formats as containers of memory – and the impossibility of return.

The vulnerability of physical sound carriers – and the rapidity with which they are relegated to wastelands – was also evidenced in a phono-archaeological project carried out by British sound and visual artist Graham Dunning at the Rea Garden in Birmingham in 2010.[80] Dunning worked on a site formerly occupied by a storage facility for gramophone records which had been abandoned for over fifty years, setting out to excavate, organise and classify the shellac fragments buried in the ground (see Figure 7 below). A second phase of the project began when Dunning asked what the collection of shards may sound like. He unsuccessfully attempted to directly play them by 'running the individual pieces over the stylus'. However, because 'the sections had only up to about 4 cm of groove to play', this proved a nearly impossible task. '[K]eep[ing] the needle in the groove by hand' was also difficult, especially in the case of severely deteriorated grooves. He subsequently tried to re-compose records out of disparate shards to hear what waste may sound like (recalling Knížák's and Marclay's experiments with broken records). Finally, he digitised the phonographic remainders with an inexpensive portable flatbed scanner to be able to explore their grooves more closely, a gesture playfully mimicking (and deforming) the work of audio engineers as they repair and salvage damaged groove paths.

In 2017, a residency at Global Forest (St. Georgen im Schwarzwald) allowed Dunning to further experiment with gramophone discs which he deliberated ruined by dissolving them in methylated spirits: the melted discs were used as a basis for square paintings, each of them visually different (notably in terms of colour and viscosity) though they were derived from exactly the same method. This experimentally confirmed that disc compositions varied depending on manufacturer and era of manufacturing, while recalling earlier uses of shellac as painting material. Dunning's work demonstrates a curiosity for the surface of records – and their potential reversibility (from sonic to visual media and back again). His formative project at Rea Garden became the basis for an exploratory series of works investigating records as

80   This project was entitled 'Visitor Centre: An excavation of sound'.

Figure 7: Shellac shards, Rea Garden, Birmingham, 2010 © Graham Dunning.

ruins. Some later shellac-related projects involved sampling discarded 78s (notably in collaboration with Sascha Brosamer for the 2019 album *Glocken*, which recycled some of the records produced as part of Brosamer's *Repetitive Movements*), or yet again DJing with gramophones.[81]

With its revaluation of surface noise, US sound artist and phono-archaeologist Robert Millis's *Related Ephemera* album (2020) offers another route into the residual sounds and histories of shellac discs. The album, issued on vinyl, entirely consists of shellac and wax cylinder samples. Its opening track, 'Samsara' (the Sanskrit term for the cyclicality of existence) comprises of three movements and was intended by Millis as a reflection on 'sounds dying and being reborn': the last of the three parts, entitled 'lac insects', is dedicated to the life cycle of the lac bug. Contrary to what listeners may expect, it is not a field recording of the (inaudible) female insect but a reimagination of it where the artist sought to create 'a "cicada like" ambiance with a lot of densely packed extreme sounds that change volume rapidly – with very sharp highs'.[82] The section was so strident that the pressing plant rejected it several times, asking Millis to remaster it 'with lower transients and softer volumes'. What

---

81   See also the work of turntablist Naomi Kashiwagi (UK).
82   Millis, personal communication, June 2022.

guided the project was an interest in processes of remediation, from one recording format to another.

While Dorsett (and, to some extent, Dunning) experimented with the possibilities of trans-sensory remediation (converting the sonic into the visual and reciprocally), Millis's project involves shellac, wax and vinyl – three distinct yet interconnected materials of sound reproduction. It asks how pieces 'made from the aural detritus of one medium would translate into another medium'. As with *Repetitive Movements*, *Related Ephemera* superimposes mediatic layers and draws attention to their eco-mediatic and historical inseparability: the long history of shellac lies buried within the history of vinyl (see Chapter 4). Incidentally, the *Related Ephemera* album only emerged as an afterthought. It exists as a by-product of Millis's initial project to cast a 'perfect' record entirely made of shellac, an endeavour inspired by preliminary field work he undertook in India in 2012–2013 when a Fulbright Grant allowed him to obtain first-hand knowledge of shellac manufacturing processes and visit the former record-pressing plant in Dum Dum.

The reconstructive shellac project – which denoted a fetishistic attachment to the early material of recorded sound – was abandoned because Millis became uncomfortably aware of its revivalist connotations. A more reflective approach was subsequently chosen, recalling the distinction made by Boym between restorative and reflective (or critical) nostalgia.[83] While the former mode seeks to 'faithfully' reconstruct the image of the past, the latter (closely resembling Riegl's recognition of age-value) enhances the incessant labour of time, revelling in 'patina, ruins, cracks, imperfection'.[84] As well as upcycling 78s records as part of *Protoplastic Prototypes*, Darsha Hewitt has also explored the obsolescence of recording formats in her trilogy *High Fidelity Wasteland*. The second instalment, entitled *High Fidelity Wasteland II: The Protoplastic Groove*, was a sound installation based on the replaying of shellac records at a slower speed than they were initially intended for and overemphasising the moans of their disintegrating surfaces.

In the second half of the twentieth century and early twenty-first century, reflecting the ever-accelerating technologisation of everyday life, a considerable number of artists have engaged with obsolete media materialities, emphasising in some cases their engulfing presence. The large-scale 'archives of obsolescence' devised by US artist Dan Peterman, for instance, gather 'thousands of LPs from a defunct record shop, thousands

---

83   See Boym (2001), xviii.
84   Ibid., 4.

of art and architecture texts from a defunct book shop, thousands of slides from digitising Art History departments'.[85] Peterman's accumulations elicit a physical sense of oppression: the past becomes totalised as a solid, continuous 'wall' of commodities not dissimilar from a well-stocked super-market aisle. It becomes unreal and meaningless, opaque and impenetrable. The 'archives of obsolescence' are offered as an uncracked monument – hyperbolising the monumental retro-fetishism of contemporary pop culture and aligning with Boym's definition of restorative nostalgia.[86] The discrete artistic approaches outlined above, by contrast, pay attention to the processes of ruination and decomposition by which the technological past gets continuously realised in the present. In doing so, they privilege an interstitial, infra-sonic and fragmented reading of (media) history – one which interrogates explicit cultural inscriptions as well as the 'unmarked' spaces between the grooves.

Bibliography

Adamson, Natalie and Steven Harris. 2017. 'Material Imagination: Art in Europe, 1946–72', In *Material Imagination: Art in Europe, 1946–72*, eds. Natalie Adamson and Steven Harris, 8–21. Chichester: Wiley Blackwell.

Adorno, Theodor W. 1990. 'The Form of the Phonograph Record'. Translated by Thomas Y. Levin. *October* 55: 56–61.

Baraitser, Lisa. 2017. *Enduring Time*. London, New York, Oxford, New Delhi, Sydney: Bloomsbury Academic.

Bennett, Jane. 2010. *Vibrant Matter: A Political Ecology of Things*. Durham and London: Duke University Press.

Berliner, Emile. 1888. 'The Gramophone: Etching the Human Voice'. *Journal of the Franklin Institute* CXXV (6): 426–447.

Block, René. 2018 (1989). 'Hermine Performs the Washing-Up'. In *Broken Music: Artists' Recordworks*, eds. Ursula Block and Michael Glasmeier, 6–7. New York: Primary Information.

Block, Ursula. 2018 (1989). 'Broken Music or His Master's Voice'. In *Broken Music: Artists' Recordworks*, eds. Ursula Block and Michael Glasmeier, 9. New York: Primary Information.

Block, Ursula and Michael Glasmeier, eds. 2018 (1989). *Broken Music: Artists' Recordworks*. New York: Primary Information.

---

85   Brown (2015), 25–26.
86   On retro-consumption, see Reynolds (2011).

Bourriaud, Nicolas. 2017. *L'exforme*. Paris: PUF.

Boym, Svetlana. 2001. *The Future of Nostalgia*. New York: Basic Books.

Boym, Svetlana. 2010. 'The Off-Modern Mirror'. *e-flux journal* 19: 1–9.

Brown, Bill. 2015. 'Prelude: The Obsolescence of the Human'. In *Cultures of Obsolescence: History, Materiality and the Digital Age*, eds. Babette B. Tischleder and Sarah Wasserman, 19–38. New York: Palgrave Macmillan.

Dagognet, François. 1997. *Des détritus, des déchets, de l'abject: Une philosophie écologique*. Le Pressis-Robinson: Institut Synthélabo.

Dorsett, Chris. 2020. 'Voice Over: Archived Narratives and Silent Heirlooms'. *entanglements* 3 (1): 43–54.

Estep, Jan and Christian Marclay, 2014. *On & by Christian Marclay*. London: Whitechapel Gallery.

Flusfeder, David. 2015. 'Vinyl Road Trip'. *Granta* 138: 29–57.

Forty, Adrian. 1992 (1986). *Objects of Desire: Design and Society since 1750*. London: Thames & Hudson.

Groys, Boris. 2014. *On the New*. Translated by G. M. Goshgarian. London, New York: Verso.

Harris, Mark. 2017. 'Turntable Materialities'. *Seismograf* [online]. https://seismograf. org/fokus/sound-art-matters/turntable-materialities.

Ingold, Tim. 2012. 'Towards an Ecology of Materials'. *The Annual Review of Anthropology* 41: 427–442.

Iversen, Margaret. 1993. *Alois Riegl: Art History and Theory*. Cambridge, Massachusetts and London, England: The MIT Press.

Jørgensen, Kenneth Mølbjerg and Anete M. Camille Strand, 2014. 'Material Storytelling Learning as Intra-Active Becoming'. In *Critical Narrative Inquiry: Storytelling, Sustainability and Power*, eds. Kenneth Mølbjerg Jørgensen and Carlos Largacha-Martinez, 53–71. Hauppauge: Nova Publishers.

Leslie, Esther. 2020. 'Devices and the Designs on Us: Of Dust and Gadgets'. *West 86th: A Journal of Decorative Arts, Design History, and Material Culture* 27 (1): 3–21.

Marks, Laura U. 2000. *The Skin of the Film: Intercultural Cinema, Embodiment and the Senses*. Durham and London: Duke University Press.

Panopoulos, Panayotis. 2018. 'Vocal Letters: A Migrant's Family Records from the 1950s and the Phonographic Production and Reproduction of Memory'. *entanglements* 1(2): 30–51.

Parikka, Jussi. 2015. *A Geology of Media*. Minneapolis and London: University of Minnesota Press.

Rabaté, Jean-Michel. 2018. *Rust*. New York, London, Oxford, New Delhi, Sydney: Bloomsbury Academic.

Ratréma, Béatrice. 2020. 'Vinyl, disques et pochettes d'artistes, la collection Guy Schraenen'. *Volume* 7 (2): 211–216.

Reynolds, Simon. 2011. *Retromania: Pop Culture's Addiction to Its Own Past*. London: Faber & Faber.

Riegl, Alois. 1996. 'The Modern Cult of Monuments: Its Essence and Its Development'. In *Readings in Conservation: Historical and Philosophical Issues in the Conservation of Cultural Heritage*, eds. Nicholas S. Price, M. Kirby Talley, Jr. and Alessandra Melucco Vaccaro, 69–83. Los Angeles: Getty Publications.

Roy, Elodie A. 2016 (2015). *Media, Materiality and Memory: Grounding the Groove*. London and New York: Routledge.

Schnapp, Alain. 2018. 'What Is a Ruin? The Western Definition'. *Know: A Journal on the Formation of Knowledge* 2 (1): 155–173.

Sorensen, Diana. 2018. 'Mobility and Material Culture: A Case Study'. In *Territories and Trajectories: Cultures in Circulation*, ed. Diana Sorensen, 151–160. Durham and London: Duke University Press.

Stewart, Susan. 2007 (1993). *On Longing: Narratives of the Miniature, the Gigantic, the Souvenir, the Collection*. Durham and London: Duke University Press.

Thill, Brian. 2015. *Waste*. New York, London, Oxford, New Delhi, Sydney: Bloomsbury.

Conclusion: Sonic sculptures

Abstract:
The conclusion briefly revisits the main findings of the book and reasserts the benefits of a combinatory framework for the study of media artefacts. It also tentatively outlines a new approach to the study of phonography, moving from the spectral (or hauntological) conceptual framework (with its insistence on the trace and sonic inscriptibility) to an embodied tactile framework.

Keywords: phonography, hauntology, senses, memory, labour, embodiment

In his poem 'New Year's Letter (1995)', Canadian poet Craig Poile writes about the time-dilating experience of painting a friend's home, attentively coating the worn and damaged walls with fresh layers of paint to give them a new lease of life. The poem establishes a parallel between the exteriorised process of painting and the alterations affecting the narrator's mind, as he reflects on time past and the shedding of former selves. Both outer and inner changes are carefully recorded in the medium of the poem, 'stroke by stroke'.[1] 'New Year's Letter' evokes the erasure of past selves and moments, the taking leave of old friends and the emergence of novel relations. The final step, the poet notes, is to varnish the walls with a thin layer of shellac. "'Shellac!'", he exclaims in the last stanza, 'seals in the icons as I leave, turning from / the traces of hands going dull on the wall, / the tactile moment where a life takes shape'.[2] The scene is momentarily sealed, yet not completely arrested, and the last word is one of quiet wonder and suspension – an intimation of the unknown, unspent and yet unlived shape of things to come. Poile leaves the future open: the final coat of shellac is still drying as the poet departs.

Poile's poem about painting reminds me of what material culture theorist Thomas J. Schlereth once wrote about the practice and demands of historical

---

1   Poile (1998), 26.
2   Ibid., 27.

Roy, E.A., *Shellac in Visual and Sonic Culture: Unsettled Matter*. Amsterdam: Amsterdam University Press, 2023
DOI 10.5117/9789463729543_CON

writing, comparing it to the carpenter's craft.[3] There is something patient, repetitive, at once technical and intuitive, abstract and embodied, about building a text – about seeking to assemble something which, though inevitably imperfect, will work and stand by itself (at least for a little while). Schlereth, like Perec and his wonderfully eclectic novels (particularly *Life: A User's Manual*), understood writing as both producing and expanding space. Both believed that words could affect material environments and alter our perception of everyday objects and spaces. Perhaps it is inevitable that writing about objects and materials – and thinking with them, as I do with shellac – should make us question with renewed intensity the particular relationship between words and things. I find the image of the text as a house particularly illuminating. Although a text is not a house, both may be seen as dwelling places: in hermeneutic practice, interpretation may be understood as the patient exploration of 'the house of text'.[4] This exploration never ends: it is simultaneously an act of production; the house of text keeps expanding, with new rooms ceaselessly appearing as we go along. Another room may always be found or re-entered from a different door. And there are places, too, we may not be able to penetrate, thresholds we may not cross, silences that are so thick and settled they may never be bridged, voices that will not be released. Houses, like texts – be they poems or scholarly works –, are never quite finished. But they have to end somewhere. This conclusion draws together the significance of the research presented in the preceding chapters, while maintaining what I consider to be a productive sense of non-closure and indeterminacy.

My argument has proceeded cumulatively and iteratively. In doing so, it has reflected the discursive and sensory hybridity of shellac, and progressively defamiliarised the historical media artefact of the gramophone record by bringing attention to its pre-mediatic history and afterlives (notably, but not only, in art and design practices). One of the main suggestions of this book is that temporary moments of defamiliarisation and disorientation serve important and durable heuristic functions, without which no novel insights may emerge. In doing so, the book implicitly challenges some of the expectations and codes of contemporary academic publishing. Indeed, the current trend in scholarly writing is to provide easily consumable 'takeaway points' or 'bite-sized' chunks of information: there exists a pressure to read – but also to think and live – quickly, to hurry, to get things done (no matter how or why). Time is forever slipping through our fingers. In many

3   Schlereth (1992), 422.
4   See Dibadj (1998), n.p.

ways, the pressures and expectations of academic writing mirror those of our techno-liberal world. Society makes certain demands: it demands that we are productive, efficient, responsive, adaptable, unhesitant. It demands that we waste no time. It demands resolution rather than exploration and experimentation. There is no dead time in digital life. The smartphone has become an emblem of this feverishly rationalistic regime of hypercon-nectivity and hyperactivity. In such a context, slowness, contemplation and deliberation are frequently construed as signs of weakness. But I continue to believe, as I hope this book evidences, that the value of thinking lies in thinking itself – in experimenting with all the nuances of thought – a practice which I see as necessarily collective, open-ended, quotidian and polyvocal. It follows that not every conversation needs to end neatly or unequivocally: if some angles remain a bit rough and some edges are frayed, all the better. These imperfections remind us that there is always more to be done; that there is still room to move and begin again. The value of thinking may be lying too in its chronicity and its everydayness – a chronicity which, in the words of Lisa Baraitser, is 'itself the only condition for newness, where newness is neither breach, rupture or flash, but a quiet noticing that something remains, which is the permanent capacity to begin again'.[5]

I have proposed earlier in this study that the role of media-material theorists may be to channel the power of objects to 'witness history':[6] channelling is an altogether different process from 'producing' or 'extract-ing' knowledge. Accordingly, when we channel and mediate we may also be repairing or reconstructing threads, working against disintegration and erasure. The conscious process of channelling differs too from the non-interventionist fantasy of the 'camera eye' harboured by Christopher Isherwood in the late 1930s (and examined in Chapter 3). To channel is always already to intervene. As demonstrated by the recent work of Lisa Baraitser, the role of theory may not systematically be to produce completely new knowledge but, by actively recovering the ground of the past (including outmoded discourses, practices and objects), to provide firmer and localised understandings of the present-at-hand – to carve a discrete space for dif-ferential understandings to emerge.[7] As such, it carries with it a political dimension which goes beyond narrower discussions of politics *stricto sensu*. There is value in undertaking peripheral or infra-ordinary explorations, in going against the grain, in waiting, in looking back, in navigating dead

5   Baraitser (2017), 188.
6   Marks (2000), 85.
7   Baraitser (2017).

times and dead spaces, in getting lost and found again. Defamiliarisation may constitute, importantly, a moment of heightened engagement with our surroundings – a striving to take notice, to pay attention, to acknowledge that things are not what they are or (perhaps) what we thought they were, and could be otherwise. To 'reveal' – in Cubitt's words – 'the forces operating and the possibility of their working otherwise'[8] requires imagination (more importantly, it requires that we trust and cultivate the imaginative faculty, that we take it seriously). In this sense, defamiliarisation may offer itself as a counterpoint to alienation and its deleteriously paralysing and isolating effects (so acutely felt in contemporary digital society). Ultimately, defamiliarisation may even be a first step toward forming what sociologist of digital modernity Hartmut Rosa presciently terms a relation of 'resonance' with the world (understood as an embodied and sensitive relationship to our lived environment).[9]

Some sections of this book – and an entire chapter (Chapter 5) – have been explicitly devoted to sensory modes of knowledge within the context of creative practice. I believe this is an important aspect. With its recognition of the material world, creative practice often explicitly seeks to provoke or magnify defamiliarisation – not for the mere sake of provocation, but as a means of making visible the invisible, of imagining the unimaginable, and of channelling alternative questions about the world. This book has punctually turned to artistic modes of thinking and recovered the buried philosophical insights of Dagognet on materiology – not to suggest that conventional critical thought has reached a dead end or a stasis, but because there may be places where materials go and where the beam of the intellect alone (no matter how steady) simply cannot follow. Rather, the full body – with all of the five senses – needs to be continuously involved in the process of elucidation.[10]

This book has channelled the material history of shellac, a pivotal substance of past and present media cultures, through a wide array of formerly unexamined historical and cultural sources. In doing so, it provides a solid empirical basis for the study of shellac while outlining possible future directions for the field of media-material theory. In what follows, I

8   Ibid., 5.
9   See Rosa (2019).
10   In this respect, as already noted in the introduction of this book, I dissociate myself from media archaeology's methodological insistence on distance and estrangement as modes of understanding media, and my work aligns more closely with the multidimensional, phenom-enological approach developed the past decades by film and media theorists such as Laura U. Marks and Vivian Sobchack.

summarise and readdress the methodological implications of a combinatory material-driven approach for the study of media cultures in light of the book's findings. I subsequently examine more closely the notion of adhesive media in connection to historiographical practice. Lastly, the final section of my conclusion asks what a 'tactile moment' might mean for the study of recorded sound, pointing to a new theoretical direction. What could a shift of orientation from 'inscriptibility' to 'tactility' entail in the context of sound studies?

Situated findings

This book has mapped out a selection of moments in the long history of shellac. I have been concerned with the environmental history of the gramophone disc as well as the environments shellac has generated and passed through. Accordingly, the book has followed the material from its pre-sonic visual uses and understandings in South Asia and Europe to its transformation into a full-fledged medium of sound in the late nineteenth century and its 'golden age' in the interwar period. Its toxic material and ideological recuperation during the two World Wars was recovered, as well as its visual and sonic remediations in contemporary art and design. The different chapters served as narrative frames or containers. Each of them attended to a particular phase (corresponding to historically specific moments of extraction, inscription and expression),

Chapter 1, drawing from archival accounts and historical sources, recontextualised the early imaginaries and uses of shellac in South Asia, acknowledging how it mythopoetically connected – through the body of the parasitic female insect – with the realm of femininity. It later moved on to describe how lac and shellac were appropriated in the context of seventeenth-century global expansionism and became important sources of speculation. As well as showing how they were physically displaced, the chapter evidenced that this geographical and economical displacement of shellac also corresponded to a symbolic displacement, with new uses of the material consolidating. By the second half of the nineteenth century, as lac dyes fell into commercial desuetude, shellac became prized for its plasticity in industrial countries of the west and particularly in the US, where US photographer Samuel Peck patented his photographic case in 1854 (both a media container and the first shellac-moulded everyday item to be patented).

This historical approach to the material was pursued in Chapter 2, where I attended to an iconic plastic media container, the gramophone disc, which promptly displaced most of the previously patented shellac-based objects. As well as giving an account of Berliner's insights into the substance – attending jointly to his actual experimentation and to his material imagination – the chapter demonstrated that there were links to be found between pre-mediatic cultures of shellac and sonic cultures of the turn of the twentieth century. Not only were there symbolic threads uniting the two – and notably the trope of the disc as a seal of the human voice, recalling previous uses of shellac as seal, or the continued relevance of the visual trope within the acoustic – but also the transnational recording industry was inseparably bound with the colonial trade networks of the seventeenth and eighteenth century. The chapter offers a reflection on labour practices in the early twentieth-century phonographic industry, specifically recovering the racialised and feminised work which underpinned it.

While Chapter 1 and Chapter 2 were both concerned with themes of exploitation and colonial extraction, Chapter 3 engaged explicitly with the theme of phonographic consumption and expression. It highlighted the European interwar period – a moment of intensified gramophone and shellac consumption – as a vibrant intersensory moment in the life of the material, drawing from the multidimensional material culture theories and work of artists including Bauhaus instructor and multimedia practitioner Moholy-Nagy. In this central chapter of the book, the visual and sonic dimension of shellac were examined conjointly. A key object and concept here was that of the mirror, and the chapter recovers the buried link between mirrors and gramophone discs to propose an expanded reading of phonography as an art of the specular (rather than merely of the spectral).

Chapter 4 operated another transition and displacement, showing how the interwar expressivity of shellac was also bound with a logic of destructivity and toxicity. While Chapter 3 temporarily moved away from the substrate of the disc to interrogate its inscribed surface, Chapter 4 reengaged with the concrete materiality of shellac. It gave an account of how both the media-material resource as well as recorded sound more generally were materially and ideologically repurposed in the context of the two World Wars, frequently serving lethal and barbaric ends. Malabou's theses on destructive plasticity were discussed to provide a complementary reading of shellac and evidence its material and symbolic mutations. In doing so, the chapter argued for a widened ontologisation of the materials of culture. An account of how the resource became economically and culturally disaffected in the aftermath of the Second World War was also outlined, demonstrating how vinyl (which came to replace the material as a key phonographic supply) was strategically developed in the US in response to the war economy and to the increased geopolitical remoteness of India (the ubiquitous adoption of vinyl in western media cultures in the aftermath of the Partition is no historical coincidence).

Chapter 5 continued to chart the transformative materiality of shellac in the context of its sociocultural and mediatic twilight. It engaged with the work of contemporary art and designers, and notably interrogated the cultural legacies of gramophone art in the early twenty-first century. This last chapter also punctually reconnected with themes and patterns previously discussed in the book – and most closely reengaged with the intersensoriality and intermediality of shellac as already posited in interwar art and design cultures. In doing so, it reiterated the significance of embodied modes of thinking for the theorisation of materiality.

The different moments which are retraced in this book are not autonomous periods: despite their obvious differences, it may be suggested that they belong together at an intimate, material level. Taken together, they constitute the various strata of an ever-changing material history where each layer may endlessly recall, readdress, or anticipate, other layers. This explains why the same themes, objects, surfaces and images keep recurring through the pages – why, for instance, hands, faces, and mirrors should reappear with such pressing insistence, in altered yet recognisable forms. Throughout, the intersensory and intermedial nature of shellac has been highlighted and analysed in relation to a number of material practices and artefacts.

A material-driven approach recognises that natural resources and energies – as well as cultural and epistemic formations including infrastructures – may be transformed but that they never simply vanish: rather, they live on as surplus or residual traces (in the sense of the Foucauldian archive), latently informing the contemporary matrix. An important goal of this book was therefore to evidence, notably through archival research, the colonial unconscious stirring underneath the transnational phonographic industry at the turn of the twentieth century, and to recover some of the many histories (including histories of labour, exploitation, displacement and destruction as well as of expression, creativity and love) contained within a familiar commodity. A deceptively familiar and innocuous object, the 78s gramophone record simultaneously conceals and crystallises a variety of fine mediations, (micro-)materialities, activities, socialities, localities, histories, and temporalities. The transnational nature – and implicit colonial violence – of early phonography is latent though materially retraceable, starting with the raw materials of the record, even before the record got engraved or inscribed. Evidencing these hidden histories was done through addressing the materialised deep time of the gramophone disc – and examining both its substrate and its surface.

The present study therefore began with the apparently blank or uninscribed record – a record which, however, is not 'meaningless' or silent for it may always already contain a larger ideological notation. There survives a romantic imagination of sound as that which continuously exceeds the map, unfolding freely, beyond borders. At the same time, a closer look at the mundane materials of musical records – and a consideration of the territories, bodies, materialities and routines constitutive of modern audio cultures – alerts us to what could be called a para- or infra-audible dimension of musical cultures under modern capitalism. A focus on materiality may help us slowly bring to the surface the subterranean as well as subaltern voices, bodies and energies which fashioned the early record industry, allowing us to trace their lasting resonance to the present day. The history of phonography can be reread as a cycle of absorptions and metabolisations: an entombment of insects, of labouring forces (both in shellac factories and record-pressing plants), of subaltern subjects, but also of 'speaking/ singing/sounding bodies'.[11] The particular, now largely disused commodity of the gramophone record almost appears as a Benjaminian fossil – a commodity fossil in which the past became momentarily encapsulated and comprehensible, whilst escaping or deferring complete petrification. Indeed, no historical episode ever reaches completion or closure: no matter how distant, it continues to reverberate upon – and inflect – today's media-cultural condition.[12]

Despite its apparent reductionism and narrowness, I argue that a focus on discrete materials therefore enables us to consider the wider indeterminacy and uncertain ontology of media objects across long segments of time and space – as well as their porous boundaries. A material-driven approach – which pays attention to circulation and critical moments of transformation and disappearance – provides a dynamic rather than a static entry point into the analysis of cultural artefacts. In Cubitt's words, '[a] necdotal interpretation begins not at the level of meanings but at the prior level of mediations – the materials, energies, and connections comprising the event'.[13] As this book has demonstrated, media artefacts are not closed circuits or stable units: they ecologically coexist and are embedded within long material and discursive chains.

11   Engh (1999), 55.
12   See Benjamin (1973), 256.
13   Cubitt (2020), 17.

From a methodological point of view, this also means that there is no fixed place to begin the enquiry but that anywhere in the chain may constitute a valid starting point, and yield different insights and perspectives. Open-endedness and instability are therefore an ingrained aspect of any material-driven approach. It must be noted that the close examination of material resources and processes does not come as a 'natural' gesture for most media theorists (media archaeologists and ecomedia scholars aside). As our everyday encounters with media suggest, the technological is often unreflectively absorbed (or incorporated) as a second nature. An artificial segmentation continues to prevail between the cultural and the natural realms as well as being between cultural and natural times (which are often pitched against one another). In such a context, it is easy to forget or overlook how closely connected these 'shapes of times' are[14] – and how much past and present media cultures derive their apparent stability from the instability not only of discrete organic and non-organic substances but also of local and geopolitical relations. The study of materials allows us to challenge (and bypass) the artificial demarcation between nature and culture, leading us to re-examine the historically and politically produced categories of archaism and modernity, raw substances and finished products, natural and techno-historical times. Looking at the political and physical 'resources of media' further invites us to interrogate matters of labour and processes of production as well as moments of consumption, obsolescence, and disappearance – and readdress the relationship between materiality and historiography.[15]

Adhesive and connective media

The plastic, adhesive and reflective properties of shellac make it an intermediary object, one which cuts across time and cultural practices and, by the same token, may evidence their connectedness. Here, the term and the material of 'shellac' have functioned as a structuring concept throughout the book. By 'following the material', as recommended by Ingold, I have tentatively mapped some of its physical and symbolic metamorphoses in the course of centuries and put them into perspective, activating unforeseen

14   See Kubler (1962).
15   I borrow the phrase 'resources of media' from Viktoria Tkaczyk and Christine von Oertzen who organised a series on lectures and seminars on media materialities at the Humboldt University of Berlin in 2021.

networks of actual and metaphorical correspondences. The different moments outlined in the book are not complete as such, or irremediably set
in stone. They must be understood as being situated and conditioned by
the fleeting and particular actuality of the present. Following a material
doesn't imply automatically following – or excavating – a set, pre-grooved
path: it is a nonlinear, inventive and recursive methodology where one
may continuously retrace their steps and revisit the same scenes, only to
measure how much they have changed in the interval.

The logic of retracing may be understood within the wider analogical
or combinatory framework developed by Barbara Maria Stafford in her
thought-provoking theses on visual analogy in the pre- and post-cartesian
realms.[16] Charting the historical decline and disintegration of analogical
thinking, Stafford has observed that we 'possess no language for talking
about resemblance, only an exaggerated awareness of difference', leading
to a disconnected understanding of our environment and of one another.[17]
Her originality was to develop a 'dialectics of reconciliation', offering us
theoretical tools to make the (digital) present intelligible.[18] It is important
to note that analogy is always a reparative mode which 'grapple[s] with
the problem of how to conjoin an accumulated body of practices to the
shifting present and elusive future'.[19] It valorises links, mediations and
connections – while acknowledging unbridgeable ruptures, discontinuities
and accidents.

As such it is both anachronistic and future-oriented. This book has
sought to reconnect different geographies, durations, objects, histories
and epistemes: to some extent it can be read and understood as a gesture of
repair, echoing shellac's adhesive properties. Sculptors mend broken plaster
casts by shellacking their shards together.[20] In such a process, the whole
is never made whole again: its joints are magnified. Analogical thinking
doesn't seek to achieve homogeneity or to force fragments into a harmonious
or comprehensive whole. Not everything fits together or can be made to
artificially belong together. Yet, by seeking connections, an analogical
methodology may establish a common ground of understanding, no matter how ruptured or ephemeral. As Stafford notes, 'all items not only are
connected forward and backward over time and space but are constantly

16   See Stafford (1999).
17   Ibid., 10.
18   Ibid., 14.
19   Ibid., 133.
20   Rich (1973 [1947]), 71.

being packaged and repackaged under the pressure of different contexts'.[21]
The analogical methodology closely relates to Cubitt's anecdotal approach
where stories and fragments, because they are 'unstably connected', 'cannot
be fixed in series'.[22] As such they do not offer a linear or coherent historical
narrative. It follows that an account of mutable (media) materials provides
a record of alterations and erasures – as much as it recognises moments of
correspondence across heterogeneous spatiotemporal sites.

   The recognition of history as a heterogeneous ground and 'momentary'
object of perception – which may only be (incompletely) seized through the
present – is not synonymous with a liquidation, trivialisation, or relativisa-
tion of the historical past. While the past may never be retrospectively
salvaged (as argued by Malabou), a gesture of acknowledgement may still be
possible – a gesture that is also a preventive, future-oriented motion where
(according to an ecocritique perspective) the forces operating in the past
are uncovered to reveal 'the possibilities of them working otherwise'.[23]

Flesh and bones: From sonic inscriptibility to tactility

Conventional histories of phonography – understood as the *writing* or *etching*
of sound – underline how the possibility to store and retrieve sound shook the
deeper conditions of perception itself, challenging the sensory and cognitive
foundations of experience – as well as allowing for a new, autonomous
architecture of memory to be built. Pierre Nora notes how externalised forms
of media storage – constituting discrete *lieux de mémoire* – differ from the
enveloping, matricial environment of the *milieux de mémoire* (associated
with regimes of collective, oral memory and tradition).[24] For Nora, the
expulsion from the *milieux de mémoire* irreversibly marks the beginning of
history, historical time and consciousness. In this understanding, history
is not compatible with memory (conceived of as lived collective practice
or unconscious tradition). Indeed, he argues that '[i]f we were able to live
within memory, we would not have needed to consecrate *lieux de mémoire*
in its name. Each gesture, down to the most everyday, would be experienced
as the ritual repetition of a timeless practice in a primordial identification
of act and meaning. With the appearance of the trace, of mediation, of

21   Stafford (1999), 171.
22   Cubitt (2020), 41.
23   Ibid., 5.
24   See Nora (1989).

distance, we are not in the realm of true memory but of history'.[25] Nora's
conceptualisation of history can be understood as memory lived *from the
outside* – memory which has become simultaneously (and paradoxically)
impersonal and self-aware through the mediation (and retrieval) of discrete
material traces. What Nora describes is the emergence of a certain form
of consciousness. The storing of phonographic data participates in an
externalised culture of the trace, and has allowed for a substantial musical
patrimony (the gendering of the term is telling) to be accumulated. This
understanding of phonography as an inscription or a memory trace, which
largely dominates theoretical discourses on recorded sound, is extremely
useful to understand how it gave rise to novel conceptions of time and
memory, as well as establishing novel relationships to the cultural past
more generally. Yet describing phonography according to the logic of inscription has certain limitations. In particular, the inscriptive trope doesn't
fully accommodate the flexible shapes of labour and epistemes informing
sonic objects – nor does it recognise the new energies they might release
as they come to pass. I see polyvalence, instability, and plasticity as key
characteristics of early phonographic cultures. If, as argued by Rothenbuhler
and Durham Peters, 'phonography is at its heart a physical process',[26] we
must reconnect it to bodies and the realm of experience in a fuller and less
episodic manner.

In what follows, I would like to suggest – as a new point of departure and
an opening – that recorded sound may be complementarily approached
as a three-dimensional, plastic or sculptured object in time. This may at
first appear to be a paradoxical claim, for the sculptural and the sonic
are habitually seen as strictly opposing one another. In his enduringly
influential *Laocoön*, for instance, Lessing differentiated between dynamic,
time-based art forms (such as poetry and music) and still, space-based ones
(such as painting and sculpture). The two types were seen as irreconcilable.
Yet there may be a third way, a place of intermediation where sound relates
to touch, with phonography lying at the suture of touching, hearing and
seeing.

The philosophical novels of Ukraine-born Brazilian writer Clarice Lispector – and their agile thematising of the sonic and the temporal – may offer us
a source of inspiration in this endeavour. In her short novel *Água Viva*, first
published in 1973, Lispector has reflected on the ambiguous materiality of
music. Described by its author as 'the story of instants that flee like fugitive

tracks seen from the window of a train', *Água Viva* is an improvisatory and polyphonic novel, an attempt to capture the slippery immediacy of experience (and of what we call, maybe without ever attaining it, the living core of the 'present').[27] Addressing an imaginary other, Lispector writes:

> I see that I've never told you how I listen to music – I gently rest my hand on the record player and my hand vibrates, sending waves through my whole body: and so I listen to the electricity of the vibrations, the last substratum of reality's realm, and the world trembles inside my hands.[28]

The body is presented as a medium conditioning 'the modes of expression and reception of affects'.[29] The above passage – which must be read in conjunction with the writer's whole body of work – further encapsulates the narrator's cosmological (and even mystical) sensibility, where musical experiences may also offer a route beyond objective reality. Lispector insists not on the individual musical composition itself but on the tiny vibrations rippling through her hands – as a way of bridging distances and tangibly composing the present moment. Listening, for Lispector, also means to be and to remain physically connected to the broader world: and her record player functions like an unbroken umbilical cord, a means of keeping 'in touch' with the outside environment. The musical epiphanies so minutely described by Lispector are predicated upon a chain of material mediators – including (but not only) the record, the record player, the electrical network and the body itself (functioning as a sort of echo chamber). This hyper-mediated condition becomes acutely palpable when, at a later stage in the novel, both the record player – and the privileged form of contact with the world it gave rise to – get broken.

In recent years, the physical, and especially tactile, dimension of media in digital culture has been reclaimed by theorists who notably interrogate the surfaces of everyday devices such as smartphones and tablets. In doing so, they insist not on the 'dematerialisation' of the digital realm but on its hyper-embodied dynamics. Esther Leslie, noting the ubiquity of touch-screens, has suggested in dispirited jest that '[w]e know the feel of those

27   Lispector (1973), 66. Lispector's literary experimentation can be read alongside with Vladimir Jankélévitch's numerous philosophical essays – roughly written at the same period – on the phenomenology of the instant. Both Lispector and Jankélévitch, independently of one another, described the experience of hearing music as a 'pure present' and a moment of complete resonance (or co-incidence) with the world – a moment which vanishes without leaving a trace.
28   Ibid., 5.
29   Martin-Juchat quoted in Citton (2017), 112; my translation.

screens better than we know the contours of our lovers' bodies'.[30] For Leslie, digital tactility may signal the disaffection of social relationships and the beginning of separation – illustrating the cruel paradox of touch without contact, connection without solidary. In a joint exploration of media theory and media art, Henning Schmidgen has recently conceptualised tactility as a 'touchstone' of modern media cultures, insightfully recovering its centrality in the historiography of modernity by discussing its importance in the intersensory writings of theorists such as Walter Benjamin, Alois Riegl and Marshall McLuhan. He urges us to remember that 'our dealings with media are never limited to single sense organs. They always concern the entire body'.[31] This is the case with contemporary as well as with historical storage media such as phonographs and gramophones.

A tactile turn in phonography may allow to address jointly its material, political and affective scales – and the interdependent poles of extraction, impression and expression outlined in Chapter 2. Artistic practice may be a partner in dialogue in this difficult endeavour. If the record wavers between sound and vision (as Chapter 3 and Chapter 5 suggest), it also exists – importantly – between sound and touch. Christian Marclay's work has brought with it a new appreciation of the intimacy and affinity existing between the body and the machine. The early phonograph, with its diaphragm and its horned mouth, was partly modelled after the human body, suggesting a mimetic interchangeability between the body of the technology and that of the recordist and listener.[32] In an installation entitled *Keller and Caruso* (2000), Marclay has explored the idea of listening with the entirety of one's body. The installation was inspired by the historical encounter which took place between Caruso and Helen Keller in Atlanta in 1915, when the deaf and blind woman listened to the voice of the tenor by precariously placing her hands upon his mouth. The evocation of Helen Keller obliquely conjured up the memory of Edison's deafness – which forced him to listen to sound recordings with and through his teeth, biting into the wooden framework of the phonograph.[33] In this form of close listening, music survives as vibrations resonating through the jawbones and skull – it is intensely felt yet not heard. In her early diaries, Susan Sontag has depicted the heightened, paradoxical physicality – or the 'flesh and bones' – of listening.[34] 'Flesh and

---

30   Leslie (2020), 9.
31   Schmidgen (2022), 3.
32   See Picker (2001), 773.
33   Ibid., 776.
34   Sontag (2008), 10.

bones' are everywhere in music, and may relate equally to carnal experiences of both pleasure and pain – from the liberating physicality of playing an instrument or singing to the physically damaging labour of producing musical commodities. There may be scope to offer an embodied history of phonography, as a counterpoint to its proliferating 'disembodied voices': this would mean rethinking it not in terms of spectres and traces but, rather, of three-dimensional statues (in Serres's sense of the term). Phonography – and the making of phonographic cultures – is not only a matter of detached and autonomous sonic inscriptions. It is also a critical matter of faces, hands, and bodies working together, as many sections of this book (notably those focusing on factory work) have made clear. Now may be the time to pay more attention to the attachments and displaced feminised 'matrices', 'mothers' and 'statues' of early phonographic cultures – and to reconnect sound studies to a broader genealogy of hearing, seeing, and touching.

Bibliography

Baraitser, Lisa. 2017. *Enduring Time.* London, New York, Oxford, New Delhi, Sydney: Bloomsbury Academic.

Benjamin, Walter. 1973. *Illuminations.* London: Fontana.

Citton, Yves. *Médiarchie.* 2017. Paris: Editions du Seuil.

Cubitt, Sean. 2020. *Anecdotal Evidence: Ecocritique from Hollywood to the Mass Image.* New York: Oxford University Press.

Dibadj, Seyed Musa. 1998. 'The Authenticity of the Text in Hermeneutics: Cultural Heritage and Contemporary Change'. Series IIA Islam; vol. 4. Available online: http://www.crvp.org/publications/Series-IIA/IIA-4-Contents.pdf.

Engh, Barbara. 1999. 'After "His Master's Voice"'. *New Formations* 38: 54–63.

Ingold, Tim. 2012. 'Towards an Ecology of Materials'. *The Annual Review of Anthropology* 41: 427–442.

Kubler, George. 1962. *The Shape of Time: Remarks on the History of Things.* New Haven and London: Yale University Press.

Leslie, Esther. 2020. 'Devices and the Designs on Us: Of Dust and Gadgets'. *West 86th: A Journal of Decorative Arts, Design History, and Material Culture* 27 (1): 3–21.

Lispector, Clarice. 2012 (1973). *Água Viva.* Translated by Stefan Tobler. New York: New Directions.

Marks, Laura U. 2000. *The Skin of the Film: Intercultural Cinema, Embodiment and the Senses.* Durham and London: Duke University Press.

Nora, Pierre. 1989. 'Between Memory and History: Les lieux de mémoire'. *Representations* 26: 7–24.

Picker, John M. 2001. 'The Victorian Aura of the Recorded Voice'. *New Literary History* 32 (3): 769–786.

Poile, Craig. 1998. 'New Year's Letter (1995)'. In *First Crack*, 26–27. Ottawa, Ontario: Carleton University Press.

Rich, Jack C. 1973 (1947). *The Materials and Methods of Sculpture*. New York: Oxford University Press.

Rosa, Hartmut. 2019. *Resonance: A Sociology of our Relationship to the World*. Hoboken, New Jersey: Wiley.

Rothenbuhler, Eric W. and John Durham Peters. 1997. 'Defining Phonography: An Experiment in Theory'. *The Musical Quarterly* 81 (2): 242–264.

Schlereth, Thomas J. 1992. *Cultural History & Material Culture: Everyday Life, Landscape, Museums*. Charlottesville and London: University Press of Virginia.

Schmidgen, Henning. 2022. *Horn, or the Counterside of Media*. Durham and London: Duke University Press.

Sontag, Susan. 2008. *Reborn: Early Diaries, 1947–1964*. London: Penguin.

Stafford, Barbara Maria. 1999. *Visual Analogy: Consciousness as the Art of Connecting*. Cambridge, Massachusetts and London, England: The MIT Press.

Bibliography

Acland, Charles R., ed. 2007. *Residual Media*. Minneapolis and London: University of Minnesota Press.

Adamson, Natalie and Steven Harris. 2017. 'Material Imagination: Art in Europe, 1946–72', In *Material Imagination: Art in Europe, 1946–72*, eds. Natalie Adamson and Steven Harris, 8–21. Chichester: Wiley Blackwell.

Adarkar, Bhalchandra P. 1945. *Report on Labour Conditions in the Shellac Industry*. Delhi: Indian Labour Investigation Committee.

Adorno, Theodor W. 1990. 'The Curves of the Needle'. Translated by Thomas Y. Levin. *October* 55: 48–55.

Adorno, Theodor W. 1990. 'The Form of the Phonograph Record'. Translated by Thomas Y. Levin. *October* 55: 56–61.

Affelder, Paul. 1947. *How to Build a Record Library: A Guide to Planned Collecting of Recorded Music*. New York: E. P. Dutton & Co. Inc.

Agamben, Giorgio. 2009. *The Signature of All Things: On Method.* Translated by Luca D'Isanto with Kevin Attell. New York: Zone Books.

Altergott, Renée. 2021. 'Une machine à gloire? Legacies of the French Inventor(s) of Sound Recording through the Ages'. *French Forum* 46 (1): 19–35.

Anderson, Christy, Anne Dunlop and Pamela H. Smith, eds. 2015. *The Matter of Art: Materials, Practices, Cultural Logics, c. 1250–1750*. Manchester: Manchester University Press.

Anishanslin, Zara. 2016. *Portrait of a Woman in Silk: Hidden Histories of the British Atlantic World.* New Haven and London: Yale University Press.

Anon. 1777. *Dictionnaire des origines, ou époques des inventions utiles, des découvertes importantes, et de l'établissement des peuples, des religions, des sectes, des hérésies, des loix, des coutumes, des modes, des dignités, des monnoies, &tc*. K-M. Paris: Jean-François Bastien.

Anon. 1835. 'The Lac Insect (Chermes lacca)'. *The Saturday Magazine* 6 (175): 116.

Anon. 1936. *The Shellac Industry*. Namkum, Ranchi: The Indian Lac Research Institute.

Anon. 1940. *How to Make Good Recordings*. New York: Audio Devices, Inc.

Anon. 1942a. 'Shellac Order Strikes Disc Production'. *Broadcasting* 22 (16): 10.

Anon. 1942b. 'Transcribing Firms Discount Effect of WPB Shellac Order'. *Broadcasting* 22 (16): 10; 53.

Anon. 1956. *Shellac*. Angelo Brothers Limited: Calcutta.

Armand, Octavio. 1994. *Refractions*. Translated by Carol Maier. New York: Lumen Books.

Asendorf, Christoph. 1993. *Batteries of Life: On the History of Things and Their Perception in Modernity*. Translated by Don Reneau. Berkeley, Los Angeles, London: University of California Press.

Ballin, Anita. 1996. 'Women's Work in the First World War'. In *Women in Industry and Technology from Prehistory to the Present Day*, eds. Amanda Devonshire and Barbara Wood, 235–241. London: Museum of London.

Barad, Karen. 2017. *Meeting the Universe Halfway: Quantum Physics and the Entanglement of Matter and Meaning*. Durham and London: Duke University Press.

Baraitser, Lisa. 2017. *Enduring Time*. London, New York, Oxford, New Delhi, Sydney: Bloomsbury Academic.

Barry, Andrew. 2005. 'Pharmaceutical Matters: The Invention of Informed Materials'. *Theory, Culture & Society* 22 (1): 51–69.

Bartels, Michiel H. and Léon M. van der Hoeven. 2005. 'Business from the Cesspit: Investigations into the Socio-Economic Network of the Van Lidth de Jeude Family (1701–78) in Tiel, the Netherlands, on the Basis of Shellac Letter-Seals from a Cesspit'. *Post-Medieval Archaeology* 39 (1): 155–171.

Barthes, Roland. 1991 (1957). 'Plastic'. In *Plastics Ages: From Modernity to Postmodernity 1960–1991*, ed. Penny Sparke, 110–111. London: V&A Publications.

Baudelaire, Charles. 1995 (1863). *The Painter of Modern Life and Other Essays*. Translated and edited by Jonathan Mayne. New York: Phaidon Press.

Bauman, Zygmunt. 2000. *Liquid Modernity*. Cambridge: Polity Press.

Baxter, Walter. 1954. *The Image and the Search*. Melbourne, Toronto, London: William Heinemann.

Belchior Caxeiro, Susana. 2021. *Immaterial in the Material: A Study on 78rpm Audio Carriers in Portuguese Collections*. Nova University Lisbon, unpublished doctoral thesis.

Bell, L. M. T. 1936. *The Making & Moulding of Plastics*. London: Hutchinson's Scientific & Technical Publications.

Benjamin, Walter. 1973. *Illuminations*. London: Fontana.

Benjamin, Walter. 2003. 'On the Concept of History'. In *Selected Writings, vol. 4, 1938–1940*, eds. Howard Eiland and Michael W. Jennings, 389–400. Cambridge, Massachusetts: Belknap Press / Harvard University Press.

Bennett, Jane. 2010. *Vibrant Matter: A Political Ecology of Things*. Durham and London: Duke University Press.

Berenbaum, May. 1995. *Bugs in the System: Insects and Their Impact on Human Affairs*. Cambridge, Massachusetts: Perseus Books Blake.

Berliner, Emile. 1888. 'The Gramophone: Etching the Human Voice'. *Journal of the Franklin Institute* CXXV (6): 426–447.

Berliner, Emile. 1895. 'Technical Notes on the Gramophone'. *Journal of the Franklin Institute* 140 (6): 419–437.

Berliner, Emile. 1913. 'The Development of the Talking Machine'. *Journal of the Franklin Institute* 176 (2): 189–200.

Berthoud, Françoise et al. 2012. *Impacts écologiques des technologies de l'information et de la communication: Les faces cachées de l'immatérialité.* Paris, EDP Sciences.

Bhabha, Homi K. 2018. 'Introduction: On Disciplines and Destinations'. In *Territories and Trajectories: Cultures in Circulation*, ed. Diana Sorensen, 1–12. Durham and London: Duke University Press.

Blake, Eric C. 2004. *Wars, Dictators and the Gramophone, 1898–1945.* York: William Sessions Limited.

Block, René. 2018 (1989). 'Hermine Performs the Washing-Up'. In *Broken Music: Artists' Recordworks*, eds. Ursula Block and Michael Glasmeier, 6–7. New York: Primary Information.

Block, Ursula. 2018 (1989). 'Broken Music or His Master's Voice'. In *Broken Music: Artists' Recordworks*, eds. Ursula Block and Michael Glasmeier, 9. New York: Primary Information.

Block, Ursula and Michael Glasmeier, eds. 2018 (1989). *Broken Music: Artists' Recordworks.* New York: Primary Information.

Böhme, Hartmut. 2014. *Fetishism and Culture: A Different Theory of Modernity.* Translated by Anna Galt. Berlin and Boston: De Gruyter.

Bolter, David Jay and Richard Grusin. 1999. *Remediation: Understanding New Media.* Cambridge, Massachusetts: The MIT Press.

Bourriaud, Nicolas. 2017. *L'exforme.* Paris: PUF.

Boym, Svetlana. 1994. *Common Places: Mythologies of Everyday Life in Russia.* Cambridge, Massachusetts and London, England: Harvard University Press.

Boym, Svetlana. 2001. *The Future of Nostalgia.* New York: Basic Books.

Boym, Svetlana. 2010. 'The Off-Modern Mirror'. *e-flux journal* 19: 1–9.

Brill, Thomas B. 1980. *Light: Its Interaction with Art and Antiquities.* New York and London: Plenum Press.

Brittain, Vera. 1986 (1933). *Testament of Youth: An Autobiographical Study of the Years 1900–1925.* London: Penguin.

Bronfman, Alejandra. 2021. 'Glittery: Unearthed Histories of Music, Mica, and Work'. In *Audible Infrastructures*: *Music, Sound, Media,* eds. Kyle Devine and Alexandrine Boudreault-Fournier, 73–90. New York and Oxford: Oxford University Press.

Broven, John. 2009. *Record Makers and Breakers: Voices of the Independent Rock'n'Roll Pioneers.* Urbana and Chicago: University of Illinois Press.

Brown, Bill. 2015. 'Prelude: The Obsolescence of the Human'. In *Cultures of Obsolescence: History, Materiality and the Digital Age*, eds. Babette B. Tischleder and Sarah Wasserman, 19–38. New York: Palgrave Macmillan.

Brown, George I. 1999. *The Big Bang: A History of Explosives*. Phoenix Mill, Thrupp, Stroud: Sutton Publishing.

Bruno, Giuliana. 2007 (2002). *Atlas of Emotion: Journeys in Art, Architecture, and Film*. New York: Verso.

Bussabarger, Robert F. and Betty D. Robins. 1968. *The Everyday Art of India*. New York: Dover Publications.

Butler, Shane. 2011. *The Matter of the Page: Essays in Search of Ancient and Medieval Authors*. Madison: The University of Wisconsin Press.

Cennini, Cennino A. 1954 (1933). *The Craftsman's Handbook: The Italian "Il libro dell' arte"*. Translated by Daniel V. Thompson. New York: Dover Publications.

Chatterton, Keble E. 2017 (1826). *The Old East Indiamen*. Project Gutenberg Ebook.

Chew, Victor K. 1981. *Talking Machines*. London: Science Museum.

Citton, Yves. *Médiarchie*. 2017. Paris: Editions du Seuil.

Clarke, Graham. 1997. *The Photograph*. Oxford, New York: Oxford University Press.

Cocteau, Jean. 1999 (1930). *Opium: Journal d'une désintoxication*. Paris: Stock.

Cook, Ian. 2004. 'Follow the Thing: Papaya'. *Antipode* 36: 242.

Crichton Smith, Iain. 1990. 'On the Golan Heights'. In *High on the Wall: A Morden Tower Anthology*, ed. Gordon Brown, 119. Newcastle upon Tyne: Morden Tower / Bloodaxe.

Crockett, Clayton. 2010. 'Foreword', *Plasticity at the Dusk of Writing: Dialectic, Destruction, Deconstruction*, by Catherine Malabou. Translated by Carolyn Shread, xi–xxv. New York: Columbia University Press.

Cubitt, Sean. 2013. 'Anecdotal Evidence'. *NECSUS_European Journal of Media Studies*. Available from https://necsus-ejms.org/anecdotal-evidence/.

Cubitt, Sean. 2017. *Finite Media*. Durham: Duke University Press.

Cubitt, Sean. 2020. *Anecdotal Evidence: Ecocritique from Hollywood to the Mass Image*. New York: Oxford University Press.

Dagognet, François. 1997. *Des détritus, des déchets, de l'abject: Une philosophie écologique*. Le Pressis-Robinson: Institut Synthélabo.

Dant, Tim. 1999. *Material Culture in the Social World: Values, Activities, Lifestyles*. Maidenhead: Oxford University Press.

Daston, Lorraine. 2004. 'Introduction: Speechless'. In *Things That Talk: Object Lessons from Art and Science*, ed. Lorraine Daston, 9–24. New York: Zone books.

David, Allison Matthews. 2015. *Fashion Victims: The Dangers of Dress Past and Present*. London: Bloomsbury.

Day, Tim. 2000. *A Century of Recorded Music: Listening to Musical History*. New Haven and London: Yale University Press.

Dearle, Denis A. 1944. *Plastic Moulding*. London, New York, Melbourne: Hutchinson's Scientific and Technical Publications.

Denning, Michael. 2015. *Noise Uprising: The Audiopolitics of a World Musical Revolution*. London, New York: Verso.

Derrida, Jacques. 1989. 'Biodegradables: Seven Diary Fragments'. Translated by Peggy Kamuf. *Critical Inquiry* 15 (4): 812–873.

Derrida, Jacques. 1994. *Spectres of Marx*. New York and London: Routledge.

Derrida, Jacques. 2010. *Copy, Archive, Signature*: *A Conversation on Photography*. Stanford, California: Stanford University Press.

Devine, Kyle and Alexandrine Boudreault-Fournier, eds. 2021. *Audible Infrastructures*: *Music, Sound, Media.* New York and Oxford: Oxford University Press.

Devine, Kyle. 2019a. *Decomposed: The Political Ecology of Music*. Cambridge, Massachussetts and London, England: The MIT Press.

Devine, Kyle. 2019b. 'Musicology without Music'. In *On Popular Music and Its Unruly Entanglements*, eds. Nick Braae and Kai Arne Hansen, 15–37. Cham: Palgrave Macmillan.

Devine, Kyle. 2013. 'Imperfect Sound Forever: Loudness Wars, Listening Formations and the History of Sound Reproduction'. *Popular Music* 32 (2).

Dibadj, Seyed Musa. 1998. 'The Authenticity of the Text in Hermeneutics: Cultural Heritage and Contemporary Change'. Series IIA Islam; vol. 4. Available online: http://www.crvp.org/publications/Series-IIA/IIA-4-Contents.pdf.

Dillard, Annie. 1976 (1974). *Pilgrim at Tinker Creek*. London: Pan Books.

Dobell, Eva. 2006. 'In a Soldiers' Hospital II: Gramophone Tunes, 1919'. In *The Penguin Book of First World War Poetry*, ed. George Walter, 208. London: Penguin.

Domínguez Rubio, Fernando. 2016. 'On the Discrepancy Between Objects and Things: An Ecological Approach'. *Journal of Material Culture* 21(1): 59–86.

Dorsett, Chris. 2020. 'Voice Over: Archived Narratives and Silent Heirlooms'. *entanglements* 3 (1): 43–54.

Draaisma, Douwe. 2000. *Metaphors of Memory: A History of Ideas about the Mind*. Cambridge: Cambridge University Press.

Eastaugh, Nicholas, Valentine Walsh, Tracey Chaplin and Ruth Siddall. 2004. *The Pigment Compendium: A Dictionary of Historical Pigments*. Amsterdam, Boston, Heidelberg, London, New York, Oxford, Paris, San Diego, San Francisco, Singapore, Sydney, Tokyo: Elsevier.

Engh, Barbara. 1999. 'After "His Master's Voice"'. *New Formations* 38: 54–63.

Ernst, Wolfgang, 2013. *Digital Memory and the Archive*. Minneapolis, London: University of Minnesota Press.

Ernst, Wolfgang. 2011. 'Media Archaeography: Method and Machine Versus History and Narrative of Media'. In *Media Archaeology: Approaches, Applications, and Implications*, eds. Erkki Huhtamo and Jussi Parikka, 239–255. Berkeley: University of California Press.

Ernst, Wolfgang. 2016. *Sonic Time Machines: Explicit Sound, Sirenic Voices, and Implicit Sonicity.* Amsterdam: Amsterdam University Press.

Estep, Jan and Christian Marclay, 2014. *On & by Christian Marclay.* London: Whitechapel Gallery.

Fauche, Hippolyte. 1863. *Une tétrade ou drame, hymne, roman et poème traduits pour la première fois du sanscrit en français.* Paris: Benjamin Duprat.

Fauser, Anegret. 2013. *Sounds of War: Music in the United States during World War II.* New York: Oxford University Press.

Flusfeder, David. 2015. 'Vinyl Road Trip'. *Granta* 138: 29–57.

Föllmer, Moritz. 2013. *Individuality and Modernity in Berlin.* Cambridge: Cambridge University Press.

Forty, Adrian. 1992 (1986). *Objects of Desire: Design and Society since 1750.* London: Thames & Hudson.

Fosler-Lussier, Danielle. 2020. *Music on the Move.* Minneapolis: University of Michigan Press.

Frey, James W. 2012. 'Prickly Pears and Pagodas: The East India Company's Failure to Establish a Cochineal Industry in Early Colonial India'. *The Historian* 74 (2): 241–266.

Frey, James W. 2019. 'The Global Moment: The Emergence of Globality, 1866–1867, and the Origins of Nineteenth-Century Globalization'. *The Historian* 81 (1): 9–56.

Fuller, Matthew. 2005. *Media Ecologies: Materialist Energies in Art and Technoculture.* Cambridge, Massachusetts: The MIT Press.

Gaisberg, Fred W. 1948 (1947). *Music on Record.* London: Robert Hale Limited.

Galloway, Alexander R. 2012. 'Plastic Reading'. *NOVEL: A Forum on Fiction* 45 (1): 10–12.

Galvan, Jill. 2010. *The Sympathetic Medium: Feminine Channeling, the Occult, and Communication Technologies, 1859–1919.* Ithaca and London: Cornell University Press.

Gauss, Stefan. 2007 (2014). 'Listening to the Horn: On the Cultural History of the Phonograph and the Gramophone'. In *Sounds of Modern History: Auditory Cultures in 19th- and 20th-Century Europe,* ed. Daniel Morat, 71–100. New York, Oxford: Berghahn Books.

Gettens, Rutherford J. and George L. Stout. 1966 [1943]. *Painting Materials: A Short Encyclopaedia.* New York: Dover Publications.

Gibson, Chris and Andrew Warren. 2021. *The Guitar: Tracing the Grain Back to the Tree.* Chicago and London: The University of Chicago Press.

Gilbert, Marianne. 2017. 'Plastics Materials: Introduction and Historical Development'. In *Brydson's Plastics Materials. Eighth Edition,* ed. Marianne Gilbert, 1–18. Amsterdam, Boston, Heidelberg, London, New York, Oxford, Paris, San Diego, San Francisco, Singapore, Sydney, Tokyo: Elsevier.

Grivel, Charles. 1994 (1992). 'La bouche cornue du phonographe', translated by Douglas Kahn and Gregory Whitehead. In *Wireless Imagination: Sound, Radio and the Avant-Garde*, ed. Douglas Kahn and Gregory Whitehead, 51–75. Cambridge, Massachusetts and London, England: The MIT Press.

Gronow, Jukka. 2003. *Caviar with Champagne: Common Luxury and the Ideals of the Good Life in Stalin's Russia*. Oxford: Berg. Ebook.

Gronow, Pekka and Ilpo Saunio. 1999. *An International History of the Recording Industry*. London and New York: Cassell.

Groys, Boris. 2014. *On the New*. Translated by G. M. Goshgarian. London, New York: Verso.

Gumbrecht, Hans Ulrich. 2013. *After 1945: Latency as Origin of the Present*. Stanford, California: Stanford University Press.

Hailey, Christopher. 1994. 'Rethinking Sound: Music and Radio in Weimar Germany'. In *Music and Performance during the Weimar Republic*, ed. Bryan Gilliam, 13–36. Cambridge, New York, Melbourne: Cambridge University Press.

Hake, Sabine. 1994. 'Urban Spectacle in Walter Ruttmann's *Berlin, Symphony of the Big City*'. In *Dancing on the Volcano: Essays on the Culture of the Weimar Republic*, ed. Thomas W. Kniesche and Stephen Brockmann, 127–142. Columbia: Camden House.

Hargitt, George T. 1923. 'Invertebrate Animals and Civilization'. *The Scientific Monthly* 16 (6): 608–622.

Harris, Mark. 2017. 'Turntable Materialities'. *Seismograf* [online]. https://seismograf. org/fokus/sound-art-matters/turntable-materialities.

Heath et al. 2000. *300 Years of Industrial Design: Function, Form, Technique 1700–2000*. New York: Watson-Guptill Publications.

Herod, Andrew. 2011. *Scale*. London and New York: Routledge.

Herzogenrath, Bernd. 2015. 'Media|Matter: An Introduction'. In *Media|Matter: The Materiality of Media|Matter as Medium*, ed. Bernd Herzogenrath, 1–16. New York: Bloomsbury Academic.

Hicks, Edward. 1961. *Shellac: Its Origin and Applications*. New York: Chemical Publishing Co., Inc.

Hughbanks, Leroy. 1945. *Talking Wax or the Story of the Phonograph*. New York: The Hobson Book Press.

Ingold, Tim. 2010. 'The Textility of Making'. *Cambridge Journal of Economics* 34: 91–102.

Ingold, Tim. 2012. 'Towards an Ecology of Materials'. *The Annual Review of Anthropology* 41: 427–442.

Isherwood, Christopher. 1998 (1939). *Goodbye to Berlin*. London: Vintage.

Iversen, Margaret. 1993. *Alois Riegl: Art History and Theory*. Cambridge, Massachusetts and London, England: The MIT Press.

Jones, Geoffrey. 1985. 'The Gramophone Company: An Anglo-American Multinational, 1898–1931'. *The Business History Review* 59 (1): 76–100.

Jørgensen, Kenneth Mølbjerg and Anete M. Camille Strand, 2014. 'Material Storytelling Learning as Intra-Active Becoming'. In *Critical Narrative Inquiry: Storytelling, Sustainability and Power*, eds. Kenneth Mølbjerg Jørgensen and Carlos Largacha-Martinez, 53–71. Hauppauge: Nova Publishers.

Kaes, Anton, Martin Jay, Edward Dimendberg, eds. 1995. *The Weimar Republic Sourcebook*. Berkeley, Los Angeles, London: University of California Press.

Kasson, John F. 1999 (1976). *Civilizing the Machine: Technology and Republican Values in America, 1776–1900*. New York: Hill and Wang.

Katz, Sylvia. 1985. *Classic Plastics: From Bakelite ... to High-Tech*. London: Thames and Hudson.

Klee, Paul. 1965. *The Diaries of Paul Klee, 1898–1918*. London: Peter Owen.

Knaggs, Nelson S. 1947. *Adventures in Man's First Plastic: The Romance of Natural Waxes*. New York: Reinhold Publishing Corporation.

Kouwenhoven, John A. 1982. 'American Culture: Words or Things?'. In *Material Culture Studies in America*, ed. Thomas J. Schlereth, 79–92. Nashville: American Association for State and Local History Press.

Kracauer, Siegfried. 1966 (1947). *From Caligari to Hitler: A Psychological History of the German Film*. New York: Princeton University Press.

Krämer, Sybille. 2015. *Medium, Messenger, Transmission: An Approach to Media Philosophy*. Amsterdam: Amsterdam University Press.

Kubler, George. 1962. *The Shape of Time: Remarks on the History of Things*. New Haven and London: Yale University Press.

Kumar De, Santosh. 1990. *Gramophone in India: A Brief History*. Calcutta: Uttisthata Press.

Lacan, Jacques. 2006. Ecrits: *The First Complete Edition in English*. Translated by Bruce Fink. New York and London: W. W. Norton & Company.

Lange, Britta. 2015. '*Posterestante*, and Messages in Bottles: Sound Recordings of Indian Prisoners in the First World War'. *Social Dynamics* 41 (1): 84–100.

Lefebvre, Henri. 2013. *Rhythmanalysis: Space, Time and Everyday Life*. London, New York, Oxford, New Delhi, Sydney: Bloomsbury Academic.

Lehmann, Ann-Sophie. 2015. 'The Matter of the Medium: Some Tools for an Art-Theoretical Interpretation of Materials'. In *The Matter of Art: Materials, Practices, Cultural Logics, c. 1250–1750*, eds. Christy Anderson, Anne Dunlop and Pamela H. Smith, 21–41. Manchester: Manchester University Press.

Lehmann, Ulrich. 2015. 'Material Culture and Materialism: The French Revolution in Wallpaper'. In *Writing Material Culture History*, eds. Anne Gerritsen and Giorgio Riello, 173–190. London, New Delhi, New York, Sidney: Bloomsbury.

LeMahieu, Daniel. 1982. 'The Gramophone: Recorded Music and the Cultivated Mind in Britain Between the Wars'. *Technology and Culture* 23 (3): 372–39.

Leslie, Esther. 2016. *Liquid Crystals*. London: Reaktion.

Leslie, Esther. 2020. 'Devices and the Designs on Us: Of Dust and Gadgets'. *West 86th: A Journal of Decorative Arts, Design History, and Material Culture* 27 (1): 3–21.

Levin, Thomas Y. 1990. 'For the Record: Adorno on Music in the Age of Its Technological Reproducibility'. *October* 55: 23–47.

Lewis, Hannah. 2015. '"The Music Has Something to Say": The Musical Revisions of *L'Atalante* (1934)'. *Journal of the American Musicological Society* 68 (3): 559–603.

Lewis, Hannah. 2020. 'The *piano mécanique* in 1930s French Cinema'. *French Screen Studies* 20 (3–4): 158–179.

Lispector, Clarice. 2012 (1973). *Água Viva*. Translated by Stefan Tobler. New York: New Directions.

Lowe, Jacques, Russell Miller and Roger Boar. 1982. *The Incredible Music Machine*. London: Quartet / Visual Arts.

Mackenzie, Compton. 1925. 'The Gramophone: Its Past, Its Present, Its Future'. *Proceedings of the Musical Association* 51: 97–119.

Maillet, Arnaud. 2009. *The Claude Glass: Use and Meaning of the Black Mirror in Western Art*. Translated by Jeff Fort. New York: Zone Books.

Maisonneuve, Sophie. 2002. 'La constitution d'une culture et d'une écoute musicale nouvelles: Le disque et ses sociabilités comme agents de changement culturel dans les années 1920 et 1930 en Grande-Bretagne'. *Revue de Musicologie* 88 (1): 43–66.

Maisonneuve, Sophie. 2006. 'De la machine parlante au disque: Une innovation technique, commerciale et culturelle'. *Vingtième Siècle. Revue d'histoire* 92 (4): 17–31.

Malabou, Catherine. 2010. *Plasticity at the Dusk of Writing: Dialectic, Destruction, Deconstruction*. Translated by Carolyn Shread. New York: Columbia University Press.

Malabou, Catherine. 2012. *Ontology of the Accident: An Essay on Destructive Plasticity*. Translated by Carolyn Shread. Cambridge, Malden, Massachusetts: Polity Press.

Mandal, Jyoti Prakash. n.d. *A Study of the Problems and Prospects of Lac Industry in the Purulia District of West Bengal*. University of Burdwan, unpublished doctoral thesis.

Marchand, Eckart. 2015. 'Material Distinctions: Plaster, Terracotta, and Wax in the Renaissance Artist's Workshop'. In *The Matter of Art: Materials, Practices, Cultural Logics, c. 1250–1750*, eds. Christy Anderson, Anne Dunlop and Pamela H. Smith, 160–179. Manchester: Manchester University Press.

Marks, Laura U. 2000. *The Skin of the Film: Intercultural Cinema, Embodiment and the Senses*. Durham and London: Duke University Press.

Martland, Peter. 1997. *Since Records Began: EMI, the First 100 Years*. London: B. T. Batsford Ltd.

Masschelein-Kleiner, Liliane. 1995. *Ancient Binding Media, Varnishes and Adhesives*. Translated by Janet Bridgland, Sue Walston and A. E. Werner. Rome: ICCROM.

Mattern, Shannon. 2017. *Code and Clay, Data and Dirt: Five Thousand Years of Urban Media*. Minneapolis and London: University of Minnesota Press.

Mawani, Renisa. 2015. 'Insects, War, Plastic Life'. In *Plastic Materialities: Politics, Legality, and Metamorphosis in the Work of Catherine Malabou*, eds. Brenna Bhandar and Jonathan Goldberg-Hiller, 159–196. Durham and London: Duke University Press.

Maxwell, Richard and Toby Miller. 2012. *Greening the Media*. Oxford: Oxford University Press.

McCormack, Ryan. 2016. 'The Colossus of Memnon and Phonography'. *Sound Studies* 2 (2): 165–187.

Melchior-Bonnet, Sabine. 1994. *Histoire du miroir*. Paris: Hachette.

Melillo, Edward D. 2014. 'Global Entomologies: Insects, Empires, and the "Synthetic Age"'. *Past and Present* 223: 233–270.

Melillo, Edward D. 2020. *The Butterfly Effect: Insects and the Making of the Modern World*. New York: Alfred A. Knopf. Ebook.

Miller, Henry. 2009 (1936). *Black Spring*. Richmond: Oneword Classics.

Miller Frank, Felicia. 1995. *The Mechanical Song: Women, Voice, and the Artificial in Nineteenth-Century French Narrative*. Stanford, California: Stanford University Press.

Moholy-Nagy, László. 1969 (1925). *Painting, Photography, Film*. Translated by Janet Seligman. London: Lund Humphries.

Mosse, George. 1991 (1990). *Fallen Soldiers: Reshaping the Memory of the World Wars*. New York: Oxford University Press.

Mukhopadhyay, Asok. 2007. *A Study of the Shellac Industry with Special Reference to West Bengal*. University of Calcutta, unpublished doctoral thesis.

Müller, Martin. 2015. 'Assemblages and Actor-Networks: Rethinking Socio-Material Power, Politics and Space'. *Geography Compass* 9 (1): 27–41.

Nakamura, Lisa. 2014. 'Indigenous Circuits: Navajo Women and the Racialization of Early Electronic Manufacture'. *American Quarterly* 66 (4): 919–941.

Naylor, Gillian. 1968. *The Bauhaus*. London: Studio Vista.

Nguyen, Thi-Phuong, Xavier Sené, Emilie le Bourg, Stéphane Bouvet. 2011. 'Determining the Composition of 78-rpm Records: Challenge or Fantasy?'. *ARSC Journal* XLII: 27–42.

Nora, Pierre. 1989. 'Between Memory and History: Les lieux de mémoire'. *Representations* 26: 7–24.

Osborne, Harold, ed. 1975. *The Oxford Companion to the Decorative Arts.* Oxford: Oxford University Press.

Osborne, Richard. 2012. *Vinyl: A History of the Analogue Record.* Farnham, Burlington: Ashgate.

Ospina-Romero, Sergio. 2019. 'Ghosts in the Machine and Other Tales around a "Marvelous Invention": Player Pianos in Latin America in the Early Twentieth Century'. *Journal of the American Musicological Society* 72 (1): 1–42.

Ovid. 1974. *Metamorphoses.* Translated by Mary M. Innes. London: Penguin Books.

Ozgen-Tuncer, Asli. 2019. 'Historiographies of Women in Early Cinema'. *NECSUS* 8 (1): 273–281.

Palmié, Stephan. 2019. 'An Episode in the History of an Acoustic Mask'. *Archives de sciences sociales des religions* 187: 127–148.

Panopoulos, Panayotis. 2018. 'Vocal Letters: A Migrant's Family Records from the 1950s and the Phonographic Production and Reproduction of Memory'. *entanglements* 1(2): 30–51.

Parikka, Jussi. 2010. *Insect Media: An Archaeology of Animals and Technology.* Minneapolis and London: University of Minnesota Press.

Parikka, Jussi. 2015. *A Geology of Media.* Minneapolis and London: University of Minnesota Press.

Parry, Ernest J. 1935. *Shellac.* London: Sir Isaac Pitman & Sons, Ltd.

Parthasarathi, Vibodh. 2008. 'Not Just Mad Englishmen and a Dog: The Colonial Tuning of "Music on Record", 1900–1908'. *Working Paper* 2: 1–31.

Pastoureau, Michel. 2011 (2008). *Noir: Histoire d'une couleur.* Paris: Editions du Seuil.

Perec, Georges. 1989. *L'infra-ordinaire.* Paris: Seuil.

Petroski, Henry. 1989. *The Pencil: A History of Design and Circumstance.* London and Boston: Faber & Faber.

Petrusich, Amanda. 2014. *Do Not Sell at Any Price: The Wild, Obsessive Hunt for the World's Rarest 78rpm Records.* New York, London, Toronto, Sydney, New Delhi: Scribner.

Picker, John M. 2001. 'The Victorian Aura of the Recorded Voice'. *New Literary History* 32 (3): 769–786.

Poile, Craig. 1998. 'New Year's Letter (1995)'. In *First Crack*, 26–27. Ottawa, Ontario: Carleton University Press.

Pollard, Anthony. 1998. *Gramophone: The First 75 Years.* London: Gramophone Publications Limited.

Poschardt, Ulf. 1998. *DJ Culture.* Translated by Shaun Whiteside. London: Quartet Books.

Rabaté, Jean-Michel. 2018. *Rust.* New York, London, Oxford, New Delhi, Sydney: Bloomsbury Academic.

Radano, Ronald and Tejumola Olaniyan, eds. 2016. *Audible Empire: Music, Global Politics, Critiques*. Durham and London: Duke University Press.

Raj, Kapil. 2007. *Relocating Modern Science: Circulation and the Construction of Knowledge in South Asia and Europe, 1650–1900*. New York: Palgrave Macmillan.

Rao, Shiva B. 1936. 'Industrial Labor in India'. *Foreign Affairs*, 14 (4): 675–684.

Ratréma, Béatrice. 2020. 'Vinyl, disques et pochettes d'artistes, la collection Guy Schraenen'. *Volume* 7 (2): 211–216.

Read, Oliver and Walter L. Welch. 1976. *From Tinfoil to Stereo: Evolution of the Phonograph. Second Edition*. Indiana: Howard W. Sams & Co.

Reynolds, Simon. 2011. *Retromania: Pop Culture's Addiction to Its Own Past*. London: Faber & Faber.

Rich, Jack C. 1973 (1947). *The Materials and Methods of Sculpture*. New York: Oxford University Press.

Riegl, Alois. 1996. 'The Modern Cult of Monuments: Its Essence and Its Development'. In *Readings in Conservation: Historical and Philosophical Issues in the Conservation of Cultural Heritage*, eds. Nicholas S. Price, M. Kirby Talley, Jr. and Alessandra Melucco Vaccaro, 69–83. Los Angeles: Getty Publications.

Rosa, Hartmut. 2019. *Resonance: A Sociology of our Relationship to the World*. Hoboken, New Jersey: Wiley.

Roth, Joseph. 2003. *What I Saw: Reports from Berlin 1920–33*. Translated by Michael Hofmann. London: Granta Books.

Rothenbuhler, Eric W. and John Durham Peters. 1997. 'Defining Phonography: An Experiment in Theory'. *The Musical Quarterly* 81 (2): 242–264.

Roy, Elodie A. 2016 (2015). *Media, Materiality and Memory: Grounding the Groove*. London and New York: Routledge.

Roy, Elodie A. 2017. 'Broken Records from Berlin: The Place of Listening in *People on Sunday* (dir. Curt and Robert Siodmak/Edgar G. Ulmer, 1929)'. *Sound Studies* 3 (1): 33–48.

Roy, Elodie A. 2018. 'Worn Grooves: Affective Connectivity, Mobility and Recorded Sound in the First World War'. *Media History* 24 (1): 26–45.

Roy, Elodie A. 2020. '"Total Trash": Recorded Music and the Logic of Waste'. *Popular Music* 39 (1): 88–107.

Roy, Elodie A. 2021a. 'Another Side of Shellac: Cultural and Natural Cycles of the Gramophone Disc'. In *Audible Infrastructures: Music, Sound, Media,* eds. Kyle Devine and Alexandrine Boudreault-Fournier, 207–226. New York and Oxford: Oxford University Press.

Roy, Elodie A. 2021b. 'The Sheen of Shellac – From Reflective Material to Self-Reflective Medium'. In *Materials, Practices, and Politics of Shine in Modern Art and Popular Culture,* eds. Antje Krause-Wahl, Petra Löffler and Änne Söll, 105–119. London, New York, Dublin: Bloomsbury.

Roy Elodie A. and Eva Moreda Rodríguez, eds. 2021. *Phonographic Encounters: Mapping Transnational Cultures of Sound, 1890–1945*. Oxon and New York: Routledge.

Rust, Brian. 1975. *Gramophone Records of the First World War: An HMV Catalogue 1914–18*. Newton Abbot: David & Charles.

Scheer, Monique. 2010. 'Captive Voices: Phonographic Recordings in the German and Austrian Prisoner-of-War Camps of World War I'. In *Doing Anthropology in Wartime and War Zones: World War I and the Cultural Sciences in Europe*, eds. Johler Reinhard, Christian Marchetti and Monique Scheer, 279–309. Bielefeld: transcript Verlag.

Schlereth, Thomas. 1992. *Cultural History & Material Culture: Everyday Life, Landscape, Museums*. Charlottesville and London: University Press of Virginia.

Schmidgen, Henning. 2022. *Horn, or the Counterside of Media*. Durham and London: Duke University Press.

Schnapp, Alain. 2018. 'What Is a Ruin? The Western Definition'. *Know: A Journal on the Formation of Knowledge* 2 (1): 155–173.

Schrader, Bärbel and Jürgen Schebera. 1988. *The "Golden" Twenties: Art and Literature in the Weimar Republic*. New Haven and London: Yale University Press.

Schulz, Bruno. 1963. *Cinnamon Shops and Other Stories*. Translated by Celina Wieniewska. London: Macgibbon & Kee.

Schwenger, Peter. 2006. *The Tears of Things: Melancholy and Physical Objects*. Minneapolis: University of Minnesota Press.

Sconce, Jeffrey. 2000. *Haunted Media: Electronic Presence from Telegraphy to Television*. Durham: Duke University Press.

Sengupta, Rakesh. 2021. 'Towards a Decolonial Media Archaeology: The Absent Archive of Screenwriting History and the Obsolete *Munshi*'. *Theory, Culture & Society* 38 (1): 3–26.

Serres, Michel. 1985. *Les cinq sens*. Paris: Grasset.

Serres, Michel. 1989 (1987). *Statues: Le second livre des fondations*. Paris: Flammarion.

Silva, João. 2016. *Entertaining Lisbon: Music, Theatre, and Modern Life in the Late 19ᵗʰ Century*. New York: Oxford University Press.

Silva, João. 2019. 'Portugal, Mechanised Entertainment, and the Second Industrial Revolution'. In *Music and the Second Industrial Revolution*, ed. Massimiliano Sala, 57–80. Turnhout: Brepols

Silvers, Michael. 2018. *Voices of Drought: The Politics of Music and Environment in Northeastern Brazil*. Urbana, Illinois: University of Illinois Press.

Sinha, Sukumar. 1966. *Census of India 1961: Handicrafts Survey Monograph on Lac Ornaments*. Calcutta: Government of India Publications.

Smart, James R. 1980. 'Emile Berliner and Nineteenth-Century Disc Recordings'. *Quarterly Journal of the Library of Congress* 37 (3–4): 422–440.

Smil, Vaclav. 2013. *Made in the USA: The Rise and Retreat of American Manufacturing*. Cambridge, Massachusetts and London, England: The MIT Press.

Smith, Jacob. 2015. *Eco-Sonic Media*. Oakland, California: University of California Press.

Sobchack, Vivian. 2004. *Carnal Thoughts: Embodiment and Moving Image Culture*. Berkeley, Los Angeles, London: University of California Press.

Sontag, Susan. 2008. *Reborn: Early Diaries, 1947–1964*. London: Penguin.

Sorensen, Diana. 2018. 'Mobility and Material Culture: A Case Study'. In *Territories and Trajectories: Cultures in Circulation*, ed. Diana Sorensen, 151–160. Durham and London: Duke University Press.

Spate, Oskar H. K. 1964 (1954). *India & Pakistan: A General and Regional Geography*. London: Methuen.

Stafford, Barbara Maria. 1999. *Visual Analogy: Consciousness as the Art of Connecting*. Cambridge, Massachusetts and London, England: The MIT Press.

Stauderman, Sarah. 2004. 'Pictorial Guide to Sound Recording Media'. In *Sound Savings: Preserving Audio Collections*, ed. Judith Matz, 29–41. Washington: Association of Research Libraries.

Steinbeck, John. 1997 (1962). *Travels with Charley*. London: Arrow Books.

Sterne, Jonathan. 2003. *The Audible Past: Cultural Origins of Sound Reproduction*. Durham and London: Duke University Press.

Stewart, Susan. 2007 (1993). *On Longing: Narratives of the Miniature, the Gigantic, the Souvenir, the Collection*. Durham and London: Duke University Press.

Taussig, Michael. 1993. *Mimesis and Alterity: A Particular History of the Senses*. New York, London: Routledge.

Thill, Brian. 2015. *Waste*. New York, London, Oxford, New Delhi, Sydney: Bloomsbury.

Thompson, Daniel V. 1956 (1936). *The Materials and Techniques of Medieval Painting*. New York: Dover Publications.

Thompson, Michael. 1979. *Rubbish Theory: The Creation and Destruction of Value*. Oxford: Oxford University Press.

Tournès, Ludovic. 2011. *Musique!: du phonographe au MP3*. Paris: Autrement.

Trachtenberg, Alan. 1979 (1965). *Brooklyn Bridge: Fact and Symbol. Second Edition*. Chicago and London: The University of Chicago Press.

Turner, Peter and Gerry Badger. 1988. *Photo Texts*. London: Travelling Light.

Vágnerová, Lucie. 2017. '"Nimble Fingers" in Electronic Music: Rethinking Sound through Neo-Colonial Labour'. *Organised Sound* 22 (2): 250–258.

Ward, Alan. 1990. *A Manual of Sound Archive Administration*. Aldershot: Gower.

Westermann, Andrea. 2013. 'The Material Politics of Vinyl: How the State, Industry and Citizens Created and Transformed West Germany's Consumer Democracy'. In *Accumulation: The Material Politics of Plastic*, eds. Jennifer Gabrys et al., 68–86. London and New York: Routledge.

Weszkalnys, Gisa. 2015. 'Geology, Potentiality, Speculation: On the Indeterminacy of First Oil'. *Cultural Anthropology* 30 (4): 611–639.

Wild, Antony. 1999. *The East India Company: Trade and Conquest from 1600*. London: HarperCollins.

Wile, Frederic W. 1926. *Emile Berliner: Maker of the Microphone*. Indianapolis: The Bobbs-Merrill Company Publishers.

Williams, Gavin. 2021. 'Shellac as Musical Plastic'. *Journal of the American Musicological Society* 74 (3): 463–500.

Wilson, Percy and George W. Webb. 1929. *Modern Gramophones and Electrical Reproducers*. London, Toronto, Melbourne and Sydney: Cassell and Company.

Zielinski, Siegfried. 2006. *Deep Time of the Media*. Translated by Gloria Custance. Cambridge, Massachusetts and London, England: The MIT Press.

Zielinski, Siegfried. 2018. 'A Media-Archaeological Postscript to the Translation of Ernst Kapp's *Elements of a Philosophy of Technology* (1877)'. In *Elements of a Philosophy of Technology: On the Evolutionary History of Culture*, by Ernst Kapp, Translated by Lauren K. Wolfe, 251–266. Minneapolis and London: University of Minnesota Press.

Zinsser, William H. 1956 (2014). *A Family History and a Brief History of Wm. Zinsser & Co.* Self-published.

Zinsser, William. 2009. *Writing Places: The Life Journey of a Writer and Teacher*. New York: HarperCollins.

Zurbrugg, Nicholas. 1999. 'Marinetti, Chopin, Stelarc and the Auratic Intensities of the Postmodern Techno-Body'. *Body and Society* 5 (2–3): 93–115.

Index

Printed and bound by CPI Group (UK) Ltd, Croydon, CR0 4YY

16/04/2025

14658437-0001